T0259970

R for Conservation and Development Projects

Chapman & Hall/CRC
The R Series

Series Editors

John M. Chambers, Department of Statistics, Stanford University, California, USA
Torsten Hothorn, Division of Biostatistics, University of Zurich, Switzerland
Duncan Temple Lang, Department of Statistics, University of California, Davis, USA
Hadley Wickham, RStudio, Boston, Massachusetts, USA

For more information about this series, please visit: https://www.crcpress.com/Chapman--Hall-CRC-The-R-Series/book-series/CRCTHERSER

R for Conservation and Development Projects

A Primer for Practitioners

Nathan Whitmore

CRC Press
Taylor & Francis Group
Boca Raton London New York

CRC Press is an imprint of the
Taylor & Francis Group, an **informa** business

A CHAPMAN & HALL BOOK

First edition published 2021

by CRC Press
6000 Broken Sound Parkway NW, Suite 300, Boca Raton, FL 33487-2742

and by CRC Press
4 Park Square, Milton Park, Abingdon, Oxon OX14 4RN

© 2021 by Taylor & Francis Group, LLC
CRC Press is an imprint of Taylor & Francis Group, an Informa business

Library of Congress Cataloging-in-Publication Data

Names: Whitmore, Nathan, author.
Title: R for conservation and development projects : a primer for practitioners / Nathan Whitmore.
Description: First edition. | Boca Raton : CRC Press, 2021. | Series: Chapman & Hall the R series | Includes bibliographical references and index.
Identifiers: LCCN 2020043692 (print) | LCCN 2020043693 (ebook) | ISBN 9780367205485 (paperback) | ISBN 9780367205492 (hardback) | ISBN 9780429262180 (ebook)
Subjects: LCSH: R (Computer program language) | Mathematical statistics--Data processing. | Conservation projects (Natural resources)--Data processing. | Sustainable development--Data processing.
Classification: LCC QA276.45.R3 W47 2021 (print) | LCC QA276.45.R3 (ebook) | DDC 519.50285/5133--dc23
LC record available at https://lccn.loc.gov/2020043692
LC ebook record available at https://lccn.loc.gov/2020043693

ISBN-13: 978-0-367-20549-2 (hbk)
ISBN-13: 978-0-367-20548-5 (pbk)
ISBN-13: 978-0-429-26218-0 (ebk)

Typeset in Computer Modern font
by KnowledgeWorks Global Ltd

Dedicated to the young development workers
and conservationists of
Papua New Guinea.

Contents

III   Modelling

14 Basic statistical concepts

Preface

The motivation for this book comes from my experiences working for a conservation organisation based in Papua New Guinea, and before that, for the Department of Conservation in New Zealand. It was while I was a ranger, working for the Department of Conservation in 2007, that I was first introduced to R. However, it wasn't until 2012, when I began working in Papua New Guinea, that I began to understand the practical value of the program. Particularly, how R could be used to help with the large amount of reporting conservation and development organisations have to do for donor funded projects (which are common in the developing world).

So it was, as part of my role in supporting national staff, that I started trying to teach R. That's when I began to encounter problems. The majority of our staff had English as a second language and limited exposure to analytical thinking. The R books available, however, were made for a western academic audience and didn't match the needs of our staff. Furthermore, the inbuilt help files in R were not easily understandable, and often overly complex examples were used, or even worse, sometimes no example was ever given. Rather than helping people access R these features were turning people off the program. The internet in the developing world, due to cost, coverage issues, and unreliability continues to be a source of frustration – as a result, help is not yet at people's finger tips. Consequently, I decided to write this book. It is built on five concepts:

1. Viewing data science as part of a greater knowledge and decision making system.

2. Giving people with a non-technical background a set of basic skills so that they start to understand the usefulness of graphing, mapping, and modelling in R.

3. Using relatable examples which are typical of activities undertaken by conservation and development organisations in the developing world.

4. Realising that this book will need to act as a reference for the major functions because plain English guides are not available.

5. The book's role is to demystify R and give people confidence to use it – it is not intended to be a comprehensive statistics book.

Acknowledgements

I would like to thank the following people who generously contributed their time to read, trial, and comment on drafts of this book: Anna Bryan, Claus Ekstrøm, Derek Ogle, Grace Nugi, James Reardon, John Lamaris, Julia Reid, Ken Aho, Peter Dillingham, Stacy Jupiter, Susan Anderson, Trudi Webster, and Verena Wimmer. I would also like to thank Elodie Van Lierde for providing the cover photograph. I am deeply indebted to my editor David Grubbs, and the efforts of the wider Taylor and Francis team in helping produce this book especially: Shashi Kumar, Ashraf Reza, and Lara Spieker.

The analyses in this book would not have been possible without the ongoing maintenance of the R language and environment by the R Development Core Team. I would also like to acknowledge the developers of the main R packages used in this book:

- AICcmodavg: Marc Mazerolle
- car: John Fox and Sanford Weisberg
- caret: Max Kuhn with Jed Wing, Steve Weston, Andre Williams, Chris Keefer, Allan Engelhardt, Tony Cooper, Zachary Mayer, Brenton Kenkel, the R Core Team, Michael Benesty, Reynald Lescarbeau, Andrew Ziem, Luca Scrucca, Yuan Tang , Can Candan, and Tyler Hunt
- DHARMa: Florian Hartig
- factoextra: Alboukadel Kassambara and Fabian Mundt
- FactoMineR: Sébastien Lê and Julie Josse and François Husson
- insight: Daniel Lüdecke and Philip Waggoner and Dominique Makowski
- janitor: Sam Firke
- lubridate: Garrett Grolemund and Hadley Wickham
- MASS: Brian Ripley, Bill Venables, Douglas Bates, Kurt Hornik, Albrecht Gebhardt, and David Firth
- nlme: Jose Pinheiro, Douglas Bates, Saikat DebRoy, Deepayan Sarkar and the R Core Team – ordinal: Rune Christensen
- rpart: Terry Therneau and Beth Atkinson
- rpart.plot: Stephen Milborrow
- sf: Edzer Pebesma
- raster: Robert J. Hijmans
- remotes: Gábor Csárdi, Hadley Wickham, Winston Chang, Jim Hester, Martin Morgan, and Dan Tenenbaum
- scales: Hadley Wickham
- tidyverse: Hadley Wickham, Mara Averick, Jennifer Bryan, Winston Chang, Lucy D'Agostino McGowan, Romain François, Garrett Grolemund, Alex Hayes, Lionel Henry, Jim Hester, Max Kuhn, Thomas Lin Pedersen, Evan Miller, Stephan Milton Bache, Kirill Müller, Jeroen Ooms, David Robinson, Dana Paige Seidel, Vitalie Spinu, Kohske Takahashi, Davis Vaughan, Claus Wilke, Kara Woo, and Hiroaki Yutani.

The code in this book was developed in R version 3.6.1 and tested for compatibility with R version 4.0.2.

1

Introduction

This book is designed for people working in conservation and development project settings. It aims to:

- Give the reader an understanding of the importance of data science in projects.

- Introduce the reader to basic statistical and data science concepts.

- Get the reader quickly coding in R with a minimum fuss using relatable examples.

- Get the reader harnessing the power of R to help solve real world problems.

1.1 What is R?

R is a language and environment for statistical computing and graphics. It was developed by Ross Ihaka and Robert Gentleman at the University of Auckland, New Zealand in the 1990s. R can run on a wide variety of operating systems including Windows, MacOS, and UNIX. R, maintained by the R Development Core Team, is in a constant state of development.

1.2 Why R?

With so many statistical programs and computer languages available why, you may ask, should I learn R? R has a number of a strengths – and these are the reasons I use and recommend it:

- It's free.

- It is well established.

- It is specifically designed for data science.

- It makes reproducibility easy.

  but most importantly,

- R users have a culture of sharing code, experiences, and supporting each other. As a result, help is never far away. In this way R is so much more than a statistical program.

1.3 Why this book?

Globally, countless projects, involving billions of dollars, are underway to alleviate poverty, improve human well-being, and safeguard our natural environment. Yet the threats to our environment have never been greater and poverty continues to prevent millions of people from accessing their most basic daily needs. Despite the ease with which data can now be collected and analysed most of these projects still don't utilise data science in any meaningful way. As a consequence, lessons remain unlearned, better ways of doing interventions remain untested, and the cost-effectiveness of investments remain unknown.

Around the world there are many thousands of organisations working everyday on projects relating to poverty, inequality, climate change, ecosystem protection, human prosperity, and peace and justice. Most of these organisations are far too small to have specialist data scientists. As a result, it is left to the practitioners on the front lines to make sense of project data. These practitioners are invariably juggling multiple tasks in arduous or outright hostile conditions. Right now, in the midst of gun fire and natural disasters, committed people – many in remote locations devoid of basic infrastructure, will be writing project reports and trying to make sense of project data.

While there are many great R books none, as-yet, fit this context. This book attempts to bridge that gap. It aims to provide conservation and development practitioners with the key must-have R skills to start harnessing the power of data science.

1.4 What are development and conservation?

In the the title of this book I use the terms 'development' and 'conservation'. Both terms are very broad – for the purpose of this book I define them loosely as:

- *Development:* a discipline focused on alleviating aspects of poverty. Poverty arises when the options for a person to live their life are severely restricted. This may include access and choices as they relate to: food, water, housing, land, education, relationships, justice, health, movement, and the ability to make a living. Often development organisations will use a 'rights-based' approach when addressing these these issues. A rights-based approach not only seeks to strengthen the ability of people to access their human rights[1] but will often seek to improve the government's ability to protect those rights.

- *Conservation:* a discipline focused on safeguarding natural features, ecological function, and species, while promoting sustainable use of the environment. Many people mistakenly believe the goal of conservation must be to exclude people from nature. This view is based on a past focus for developing human-free protected areas, but this is no longer a fair representation. These days conservation is increasingly focused on issues of sustainability – which requires the involvement of people to be successful.

Both concepts are captured across the United Nation's Sustainable Development Goals (a set of 17 goals adopted by 193 countries to achieve a better and more sustainable future). As a result many development and conservation organisations now design their projects with these Sustainable Development Goals in mind (`https://www.un.org/sustainabledevelopment`). Consequently, regardless of whether an organisation defines themselves as a development or conservation organisation they often end up undertaking similar interventions – albeit with a slightly different emphasis.

1.5 Science and decision making

The decisions we make affect things. Most of the time our decisions are personal, small and affect only ourselves: like choosing which route to take to work. But everyday decisions made by organisations affect the lives of hundreds, thousands, and even millions of people. But what forms the basis for their decision making? Is it from a religious belief, a political ideology, or a tribal allegiance? Or is it from a pattern drawn from a large set of careful observations called data? It's pretty clear that decisions based on religion, ideology, or tribalism may struggle to be accepted by an outsider who has a different set of personal allegiances. Decisions based on data have, at least, the possibility of avoiding these pitfalls.

By itself data without any kind of context is just a bunch of meaningless digits or letters. In order to give the data meaning, we need to identify an underlying pattern and relate it back to a key question that we are trying

---

[1]see `https://www.un.org/en/universal-declaration-human-rights/`

to answer. If this process sounds similar to science, don't worry, you're not mistaken – it is science – data science.

These days we recognise science as a highly successful knowledge system. Unlike other knowledge systems, science uses a general method to understand the world. This method operates by posing questions, analysing observations (data), identifying patterns, and making interpretations. In some situations the system which we are observing will be actively manipulated, a process we call an experiment. But not all science is about experimentation. A lot of science focuses on understanding systems without manipulating them.

The main way science differs from the other types of knowledge (e.g. religious knowledge and indigenous traditional knowledge) is that all knowledge in science is provisional and is capable of being disproved, revised, or reinforced based on the outcome of additional scientific study. Challenging existing knowledge in science is not only perfectly acceptable, it is expected – so long as it is backed up by evidence (see Chapter 2).

1.6   Why data science is important

With the explosion in the accessibility of personal computers over the last few decades of the 20th century advanced analytical and data storage methods, once only available to a handful of professional scientists, are now available to billions of people. As a result a new discipline, data science, has developed. Not only is data science about managing, and analysing data, but it is also about making the collection and analysis of data easily reproducible, and integrating data science into decision making.

Organisations and individuals who understand the value of data and incorporate data science into their work flows are in a position to improve their products, methods, and decision making. However, despite clear benefits of incorporating data science currently few conservation or development organisations integrate it in any systematic way. This is unfortunate because conservation, and development are to some extent crisis disciplines. Often success is dependent on finding and implementing solutions before situations worsen. Consequently, it is important that these organisations understand the patterns which defined past successes or failures and implement interventions based around the best evidence at hand.

1.6.1   Monitoring and evaluation

Increasingly, and understandably, the donors who fund conservation and development projects are asking for proof that the projects they fund are working. This has resulted in organisations being required to monitor a large number of measurements (known as 'indicators') in order to evaluate the effectiveness of their activities across the lifespan of the project. This requirement is now known as 'monitoring and evaluation' (often just shortened to 'M&E'). In the absence of automation such reporting is often repetitive, time consuming, and error prone. However, if organisations can better integrate data science into their day-to-day work these tasks can become easy.

1.6.2 Projects versus programmes

The way we use data science in development and conservation depends on whether we are talking about projects or programmes. Most organisations, but not all, recognise a difference between the two:

- A project can be thought of as something temporary: a set of activities which has to deliver a certain outcome within a set deadline. A typical project may run over a few years. Typically, the use of data science will be heavily focused on:

  - Summarising project data.
  - Monitoring and evaluation (i.e. reporting on a project's effectiveness).
  - Communicating key messages with project stakeholders (including communities, partners, government, and the wider public).

  and to a lesser extent:

  - Undertaking experimental approaches to interventions.

- A programme operates on a longer time frame than a project and may run for decades. A programme exists to achieve the longer-term goals of the organisation. Projects represent stepping stones to achieving the goals of the programme. The use of data science in a programmatic setting usually focuses around:

  - Analysing the underlying causes behind the problems the organisation seeks to solve.
  - Informing the organisation's strategies.
  - Improving the overall effectiveness of the organisation.
  - Safeguarding institutional data and knowledge.

1.6.3 Project delivery versus research projects

Sometimes people from academic backgrounds struggle to adapt to conservation and development project settings. In academic settings the approach normally revolves around research projects, not project delivery. The two are fundamentally different. Research projects apply the scientific method in order to test a set of hypotheses, so long as the research results in a conclusion it is successful – it is not reliant on the preferred hypothesis of the researcher being upheld. In research projects it is the process which is important – not the outcome. This is quite different to project delivery in which the success of the project is entirely dependent on a successful outcome. This is in part due to most conservation and development projects attempting to solve real world problems. If the project fails to reach its targets and does not deliver the outcomes it promised, not only may the organisation face issues securing funding in the future, but the problem the project was designed to address could worsen.

1.7   The goal of this book

I wrote this book to be the R reference book project-based practitioners, like myself, need. Learning R by trial and error is fine – but not if you are under the constant time pressure of a typical conservation or development project. My hope is that the example scenarios and analyses present in this book are close enough to those experienced in real life project scenarios that practitioners can quickly find, learn, and adapt the code to their purposes in a matter of hours.

I don't assume the reader has a background in statistics or data science, or anymore than a high school education. The initial chapters of the book are designed to give helpful background and context for those of you who have had little exposure to data science, evidence-based decision making, or are new to project based conservation and development. In order to facilitate the learning process all example data sets are downloadable as the package `condev` at `https://github.com/NathanWhitmore/condev` (see Chapter 6).

In no way do I presume this book will cater to all your R needs. This book is only intended as an introduction. It is not designed to be a comprehensive guide. Its job is to help you gain confidence so you can begin trialing other R packages, books, and tools on your own. Like anything new, R can seem a little scary and weird at first but hopefully you'll soon begin to feel comfortable – and begin harnessing the power of data science!

1.8   How this book is organised

This book is divided into three sections:

- **Basics:** concepts around inference and evidence, project management, data science concepts, and getting started with R.

- **First steps:** practical use of R for data handling, graphing, data wrangling, and mapping. Key functions are described in simple language for users who may have difficulty understanding the R help files and lack regular access to the internet.

- **Modelling:** basic statistical concepts, modelling in R, and reporting on conservation and development projects (with worked examples).

I don't expect you will read this book from cover to cover. Rather, I imagine you will want to quickly access certain sections or examples as you need them. As a consequence there is a degree of repetition in the book. This is to ensure you can get the examples running regardless of where you start reading.

1.9   How code is organised in this book

There are two types of code examples given in the book:

1. *Runnable examples (on a grey background).* These will result in an output if run. Generally, the output will also be shown beneath these examples on a grey background, however, if the output is a graph these will appear on a white background.

2. *Non-runnable examples (on a white background).* These are either code snippets, or templates. As they do not contain all the information required (e.g. references to data sets or initial lines of code) if they are run they will return an error.

In order to distinguish data sets, weblinks, objects, variables, and R packages from ordinary text these appear in the `typewriter` font. To help the reader quickly find R functions these will appear in bold e.g. **mean()**. Functions used in the book can be found under the heading 'function' in the index. Arguments – which are the information that are required inside the functions appear in italics e.g. *method = "lm"*.

Part I

Basics

2

Inference and Evidence

In this chapter we will explore:

- How we come to a conclusion through reasoning.

- The importance of study design.

- What makes evidence.

- What makes good data.

2.1 Inference

This book is based on the idea that data science is part of a greater knowledge and decision making system. By harnessing the power of data science, we can better understand the world around us, build knowledge, and make better decisions. But in order to do that we first need some understanding of how we can come to conclusions.

As children we build rules to understand the world around us. We do this through a process called inference – coming to a conclusion through a process of reasoning. Imagine a child trying to understand what type of animal a chicken is:

- Only birds have feathers. *[Premise 1]*

- A chicken has feathers. *[Premise 2]*

- Therefore a chicken is a bird. *[Conclusion]*

This is an example of deductive reasoning. If our concepts (premises) are true then our conclusion (a chicken is a bird) will be certain. However, most things are not certain, but at best probable. Our knowledge of the universe is only partial. Even if our premises are true it is always possible for the conclusion to be false, for example:

- All of the cows for sale I have ever seen have been brown. *[Premise 1]*

- My mother is going to buy a new cow. *[Premise 2]*

– Therefore the cow she buys will be brown. *[Conclusion]*

Clearly, despite having only ever seen brown cows there is a possibility the mother buys a non-brown cow. A better way to have expressed the conclusion is:

– All of the cows for sale I have ever seen have been brown. *[Premise 1]*

– My mother is going to buy a cow. *[Premise 2]*

– Therefore the new cow will probably be brown. *[Conclusion]*

This is an example of inductive reasoning. The conclusion can never be certain, it can only be probable. Here the person making the conclusion has only seen a sample of cows, they haven't seen all the cows for sale. They are basing their conclusion on a sample. Statistics is the science of inductive reasoning. Statistics allows us to make probable conclusions from samples.

2.2   Study design

As we have seen, in our previous example, we can use statistical inference to reach a valid conclusion. However, the quality of our inference (i.e. our reasoning) depends upon our ability to remove the possibility that the same outcome could be produced by a different explanation. For example, imagine that we are interested in finding a better fish food in a village livelihood project[1]. We might try and form a conclusion from:

- **Anecdotal evidence** e.g. Aunt Elaine says that high rice bran feed is the best fish food.

- **An observational study** e.g. Aunt Elaine's fish production in her pond went from 28.2 kg to 32.1 kg in the year when she changed the feed from a low rice bran feed to a high rice bran feed.

- **A constrained study** e.g. in a before-after study involving twelve farmers, each with one pond, on average fish production improved between years from an average of 27.4 kg to 34.2 kg when a change was made from a low rice bran feed to a high rice bran feed.

- **A manipulated experiment** e.g. in a before-after-control-impact study involving twelve farmers, each with one pond, in year 1 all fish ponds received a low rice bran feed, in year 2 a random selection of 6 fish ponds received a high bran feed while the other 6 received the low bran feed. In year 1 fish production averaged 27.4 kg, while in year 2 the ponds receiving

---

[1]this is one of the worked examples found in Chapter 18

the low rice bran feed averaged 35.4 kg, and the ponds receiving the high bran feed averaged 32.9 kg.

Assuming all four examples occurred across the same time frame we can identify some key issues with each and our ability to infer anything about the effectiveness of high rice bran fish feed.

- **Anecdotal evidence**: This is the opinion of Aunt Elaine, and there is no measurement of the possible effect. There is no evidence to show that the change in production was as a direct result of the high rice bran fish feed or that a change in production even really happened.

- **An observational study**: This time we know that fish production changed between years in Aunt Elaine's pond. We also know that this coincided with the change to a high rice bran fish feed. However, these results were one-off and could be just a result of the local conditions in her pond or a consequence of some difference between years unrelated to the type of fish food used.

- **A constrained study**: This time we know that fish production changed between years coinciding with the change to a high rice bran fish feed and the results were consistent across all ponds. While the result is not likely to be due to local pond conditions, it may still be a consequence of some difference between years unrelated to the type of fish food used.

- **A manipulated experiment**: This time we know that fish production increased in the second year. We also can infer that the increase appeared to be unrelated to a change in fish food. The increase we recorded was most likely due to an unmeasured effect which differed between the study years.

From these examples we can see that our ability to infer strengthens as we measure, replicate, and use control groups. By measuring we can independently verify an outcome and compare it to other outcomes (e.g. kg of fish produced). By replicating the test (across a number of fish ponds) we get an estimate of the typical amount of fish produced and, as we increase the number of replicates, avoid the possibility that our measurements are just a one-off or chance outcome. Finally, by using a control group – a baseline treatment[2] that is held constant across the entire study – can we truly compare the effect of the different fish feed treatments used in our experiment.

The outcome of our constrained study could make many people conclude that the high rice bran feed was better – an entirely valid, but incorrect conclusion. Only in the manipulated experiment – when a true control was used – do we come to the correct conclusion that the type of fish feed was largely

---

[2]often matching a level found in natural or everyday situations

unimportant. However, there are plenty of situations in conservation and development where control groups will not be able to be used. It is easy to imagine extreme situations where the use of control groups would be completely unethical (e.g. disaster relief and disease prevention). Additionally, nowadays, most organisations engage with communities on the basis of free, prior and informed consent. This means that if control groups are to be used in projects involving communities then those communities must be aware of them and agree to their use. From the community's perspective there may be no benefit belonging to a control group – which is entirely understandable. As a result, communities may refuse to take part. Consequently, organisations need to be sensitive to such scenarios and be willing to settle for studies designed with weaker inference.

2.3   Evidence

Evidence is information which can be used in supporting a specific pattern (or hypothesis). Evidence depends heavily on context. For example, if police investigating a bank robbery find the same amount of money hidden in a suspect's house that was stolen then that information will be used as evidence against the suspect — as the information supports the police's hypothesis. However, if the police only find a goat in the suspect's house then that information is not going to be used as evidence in the case, as it doesn't link to the police's hypothesis.

In the previous section we saw that the evidence from an anecdote, a observational study, and a constrained study all supported the hypothesis that high rice bran fish feed improves production. However, a single study, the manipulated experiment, provided evidence to the contrary. It is important to recognise that evidence from a well designed experiment with strong inference is better than a larger amount of evidence from studies with weak inference. The quality of the evidence always matters more than the quantity of evidence.

2.4   What makes good data?

As I mentioned in Chapter 1, data without the context of a greater question is meaningless. In addition to answering a question, to be useful data must also have validity and reliability.

To be **valid** it must:

- *Describe what it claims to.* Take for instance a data set which claims to be about educational levels in adults. If the people collecting the data mistakenly recorded data from children then the data will not be a valid representation of adult education levels.

To be **reliable** it must:

- *Must measure the value consistently.* Imagine a case where people are required to measure the width of edible crabs for a fisheries program. If some people measure the width of the crabs with a piece of stretchy elastic bent around the crab's shell while others measure the crabs with an unbendable metal ruler the methods will not be consistent and therefore the data will not be reliable.

Good data will describe or measure what it claims to, and do it consistently.

2.5 Recommended resources

- Dan Cryan, Sharron Shatil, and Bill Mayblin (2004) Introducing Logic: a Graphic Guide Paperback. Icon Books, Cambridge.

- Dave Robinson, and Judy Groves (2013) Introducing Philosophy: a Graphic Guide Paperback. Icon Books, Cambridge.

2.6 Summary

- Statistics allows us to make probabilistic conclusions from samples.

- Inference depends on our ability to remove the possibility of alternate explanations.

- Evidence depends on context.

- The quality of the evidence matters more than the quantity of evidence.

- Good data will measure what it claims to, and do it consistently.

3

Data integration in project management

In this chapter we will explore:

- The necessity of data science in project management.

- How adaptive management cycles work.

- What a logframe is and how it relates to project data.

- How data science can be incorporated into an adaptive management cycle.

3.1 Adaptive management cycles

Integration of data science into your overall project management is not only desirable but, in most cases, essential. Not only do most organisations want to be more effective but nowadays donors require mandatory periodic reporting on project indicators. Data integration might sound complicated but it doesn't have to be. The key is to use a cyclic process known as an adaptive management cycle. Data collection, analysis, and evidence based decision making appear at different points along a project's timeline. While there are many different types of cycles most of them can be traced back to one popularised by W. Edwards Deming in the 1950s – known as the Deming cycle.

3.2 The Deming cycle

At its most basic the original Deming cycle is a repeating sequence of four steps: 'Plan – Do – Check – Act'. The Deming cycle is based on the scientific method in which the sequence: 'Plan – Do – Check' represents the 'Hypothesis – Experiment – Evaluation' component of the scientific method. The additional Act phase incorporates the outcome of the evaluation into the development (and implementation) of future plans. As a result it becomes a cycle of continuous learning.

While the original Deming cycle is pretty effective, it lacks an essential component vital for development and conservation projects: training. In a rush to deliver on a project, training staff adequately can be overlooked – especially for technical tasks such as those involving data. This, and a failure to ensure these tasks are well practiced can hinder project progress. With this in mind we can tweak the Deming cycle to be a 5-step process (Figure 3.1), in which the phases are:

1. **Plan**: designing the project.

2. **Train**: training staff and practicing key technical tasks.

3. **Do**: carrying out the project.

4. **Check**: evaluating the project (with data science).

5. **Act**: implementing changes based on the evaluation.

3.2.1   Plan

The Plan phase is about the designing of the project. The this phase is usually composed of two distinct parts:

1. Development of a project strategy and proposal.

2. Pre-implementation planning.

3.2.1.1   Development of a project strategy and proposal

The first phase in the Plan phase usually focuses on the development of a project strategy in which the logic of the project intervention is outlined (often called the project rationale). If the project is being developed in a participatory manner this is when the consultations and workshopping of the project happens with the key stakeholders occurs (e.g. communities, government agencies, and private sector). A project strategy in conservation and development typically describes:

- The current situation.
- The overall goal for the project in relation to the situation.
- The issues that currently prevent the goal from being achieved.
- The activities which could solve or neutralise those issues (i.e. logic behind the project).
- A set of indicators to measure progress towards the goal.
- The resources (money, equipment, and staff) which would be allocated to the activities.

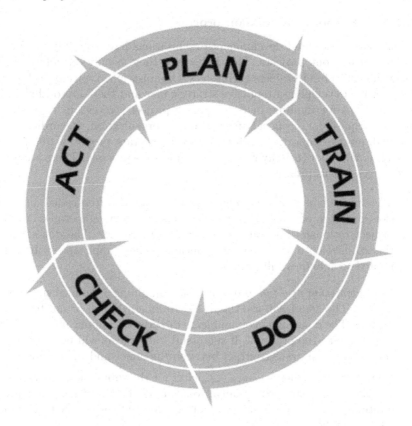

FIGURE 3.1
The project management process using a 5-step Deming cycle

Increasingly, organisations are becoming aware of the necessity for truly participatory project development with stakeholder groups who have historically often been ignored (e.g. groups representing indigenous people, women, and people with disabilities). From the point-of-view of the key stakeholders concept has been neatly summarised through the phrase:

"Nothing about us without us'

At its heart is the idea that no decision making should take place without the direct participation of the groups that would be impacted. By extension it can be applied to the way project data is consented to, collected, used, and communicated.

3.2.1.2 Proposal submission process

Most conservation and development organisations have to apply for funds in order to carry out projects. This is because they usually do not have sufficient money to fund projects themselves. Organisations typically seek project funding by: developing a tentative strategy first and then approaching donor organisations for funding, or developing a project strategy in response to a specific request for proposals from a donor organisation. Regardless of the method, most organisations will have to submit grant proposals to a prospective external donor (usually through a competitive process). There are usually two stages to this process:

1. *The submission of a summarised grant proposal* (often called an expression, or letter, of interest) in which a basic strategy (sometimes called the project rationale) and budget is given to the donor. The donor will use this to develop a short list of applicants. These applicants will then be invited to submit a full proposal.

2. *The submission of a full grant proposal* containing a detailed strategy and budget. The donor will use these to determine whether or not to fund the project. The donor may communicate directly with applicant at this stage to help refine the proposal. If successful, the process will be finalised with the signing of a grant contract between the donor and the organisation.

These days most donors have templates for the submission process. These templates can vary substantially. However, all of these will ask for a project strategy. In a full proposal you can be expected to be asked for some or all of the following:

(a) A description of your project (usually called a 'narrative') which outlines the background of your project, the issue it is attempting to solve, information about the stakeholders involved, your organisation, your strategy, and how successful outcomes of the project could be extended or reproduced in the future (usually called project sustainability).

(b) A diagram or table showing the linkage between threats, objectives, and activities (sometimes called a conceptual framework or theory of change).

(c) A table or overview showing the linkage between objectives, outputs, and activities and measurements of their effectiveness. The table form is usually called a logframe – which is short for 'logical framework', while an overview is usually called a theory-of-change.

(d) A detailed budget showing the dedication of resources to activities.

Note: From a data science perspective the logframe is the most critical. This is because the logframe outlines the type of data being collected and the criteria by which the project's success or failure will be judged. It is at this stage that the basic analytical methods for the project need to be sorted out. This is important because we need to ensure that the correct amount of financing (outlined in the proposal) will be allocated to the staff to collect, enter, and analyse the data (i.e. undertake monitoring and evaluation). Additionally, it is where we will workout our strategy to engage with external decision makers (e.g. government agencies and other institutions) so that the results of our project can inform real world decision making.

3.2.1.3   What is a logframe?

A logframe usually will outline the project's goals, outcomes, activities, and the indicators which measure their success. The logframe may also ask for the assumptions underlying the reason for the activity, as well as the physical things the project is meant to deliver (also known as deliverables), and where the data is coming from.

An example of a logframe using a 'matrix' format (based on that used by United States Agency for International Development) is given in Table 3.1. This example demonstrates the 'If-And-Then' logic which can be used to read the logframe. By itself this logic can also form the basis of a theory-of-change. We can see this if we just focus on the rows associated with the outcome and goal:

If maternal mortality decreases by 30% within 3 years [*Outcome*]

And there is no increase in other causes of mortality [*Assumption*]

Then you will have a healthy and happy community [*Goal*]

3.2.1.4   Logframe terminology

Donors often use slightly different terminology in their logframes so it is important to get advice directly from them as to what is expected in each section. Common terms used are given in Table 3.2.

3.2.1.5   Pre-implementation planning

The second phase occurs only if the proposal is successful and a contract is signed. This second phase is focused around around scheduling of staff to activities, assigning responsibilities, procurement of project equipment, and organisation of subcontracts and logistics. It is also important at this phase to document how the analysis will take place. This is especially important as some projects may last several years and the people who designed the protocols and

TABLE 3.1
An example logframe containing a theory-of-change using 'If-And-Then' logic

Level	Summary	Assumptions	Indicators	Data source
Goal	Then: Healthy and happy community	—	Before-after changes in community perception	Annual surveys of individuals
Outcome	If Maternal mortality decreases by 30% within 3 years	And There is no increase in other causes of mortality	Change in maternal mortality rate	Clinic records
Output	If 50% more women make use of health clinics during pregnancy	And Clinics can function with increased volume of clients	Change in clinic visitation rates	Clinic records
Activity	If More women are aware of maternal health services	And Women can access the clinics	Percentage of women aware of services	Survey of women
Activity	If Community maternal health workers are trained	And Participants are able to complete training	Number of people trained	Training sign-in records

wrote the proposal may not be around when all the data is finally collected and ready for analysis.

3.2.2 Train

The Train phase is often the most overlooked phase. It focuses on ensuring the staff are sufficiently trained to undertake the intervention, collect, and document the required monitoring data and, if required, undertake the final analysis. This phase should also involve active practicing or role playing if staff are relatively inexperienced, new to the survey method, analytical techniques, or technology. This phase is very important if surveys require standardised approaches, or consistent messaging. In addition, if staff are going to collect information on human subjects they should undergo specific training on how to minimise risk to participants and ensure confidentiality of personal information.

TABLE 3.2
Commonly used terms in logframes

Term	Meaning
Goal or Objective	The ultimate aim of the project (what you wish for).
Outcome or Purpose	The accumulative effect of a set of activities (what you expect to get).
Output	A promised product of the project – e.g. a positive change or the completion of something physical (like a report or some form of infrastructure). Physical outputs are sometimes called deliverables.
Indicator	What will be measured.
Data source or Means of verification	Where the data comes from for the indicators.
Activity	A specific program task or intervention whose success can be measured.
Inputs or Resources	The resources you can devote to activities.
Assumptions	The basis for your project logic.
Risks	Something which might prevent the project proceeding as planned.
Milestones	Key events or achievements showing the completion of a project stage.

3.2.3 Do

The Do phase is conceptually straightforward and involves the execution of the activity including collection of any monitoring data. Typically, this phase is the most time consuming.

3.2.4 Check

The Check phase is where the analysis of the data occurs and an evaluation of the project's effectiveness based on the measurement of the indicators outlined in the logframe. For this reason indicators have to be S.M.A.R.T :

- **S**pecific: they need to be clearly defined and understandable.
- **M**easureable: they need to be measurable from the data we will collect.
- **A**chieveable: they need to be realistic given our project's resources.

- **R**elevant: they need to link logically to the effectiveness of our activities and be able to show progress to our goal.

- **T**imely: they need to be able to give a result within our project time frame.

Note: As this book is focused on data science we will focus on quantitative indicators (i.e. measures of quantities or amounts) as indicators. However, qualitative indicators, such as people's stories, experiences and perceptions can also be used as project indicators. The use of qualitative indicators is outside the scope of this book. However, introductory material on popular approaches such as 'Most Significant Change' can be found at https://whatworks.org.nz/most-significant-change/

While this may sound surprising – conservation and development projects are often developed without assigning anyone to look after the Check phase. As a result projects can finish in a mad scramble. This happens when, in the last few days of a project, people realise that no one in their organisation has analysed, or has the skills necessary to analyse the data. As a result, the effectiveness of some projects is never properly assessed.

Many organisations use spreadsheet functions to produce their indicators. By comparison, writing code to do the same task in a program like R will be slower than using a spreadsheet method – but only the first time. Thereafter it will be faster, much faster. The results will also be less error-prone. Additionally, if an organisation is serious about improving their approaches over the long term then some form of statistical modelling will be required. Spreadsheets are not going to be enough – a statistical program like R is needed.

If you are using a programming approach (which I hope you will), the code for the analysis should have been developed in the Plan phase, practiced in the Train phase, and now it is simply a matter of running the analysis. However, it is almost inevitable that some changes to the code will be required – but these should be comparatively minor. The Check phase typically ends in the production of a report which includes:

- A review of the project's effectiveness as measured by the project's indicators.

- Evidence showing whether or not the project achieved its goals – including easy to interpret graphical information.

- A review of whether or not the project's assumptions were correct.

- Recommendations for the future direction of the project – including ways of improving project delivery and consideration of alternate strategies.

At a minimum, a copy of the report will be given to the donor. Nowadays, there is also the expectation that the report, or rather an audience appropriate version will be prepared for presentation to any participating communities, stakeholders, and key decision makers. In addition, the report, the raw data, and the analytical code should be safeguarded and archived so that the knowledge remains accessible for your organisation into the future.

3.2.5   Act

The Act phase is where the recommendations of the Check phase are implemented and decisions made. This will typically involve:

- Communicating the project's findings to external decision makers (e.g. government agencies, resource owners, and communities) to inform decision making.
- Choosing between:
  * Changing features of the project or wider programme.
  * Continuing the project without change.
  * Implementing the recommendations through a new project.
  * Terminating the project (or in severe circumstances the entire programme).

Sadly, most conservation and development projects are not designed with an Act phase. As a result, many projects end abruptly after the Check phase is completed and the donor sent a report. This is a direct consequence of most donor funded projects being designed without an adaptive management cycle in mind. There are three alternatives to this scenario:

1. Building a set of budgeted activities into the project's design to ensure an Act phase is undertaken before the project reports are submitted to the donor. In this situation a single cycle is completed.

2. Explicitly designing the project around an adaptive management cycle so that project activities are refined multiple times during the project life span. However, many donors may be resistant to taking such an approach because of their own institutional rules.

3. Resourcing the Act phase internally so it is not dependent on donor funding. In this situation the Act phase is a programme activity rather than a project activity (see Chapter 1 for a comparison between projects and programmes).

3.3 Challenges

Often in large projects staff can see themselves in isolation and not being part of a greater team process. Therefore it is important to ensure everyone understands the linkages between the project's different phases. Staff should be able to identify who is responsible for the different phases. The desire for continual improvement, however, is not universal. Indeed there are two clear instances where this desire may be lacking:

1. When an organisation has to take on a particular project for its short-term financial survival. Sometimes in these situations the project is not related to the organisation's core purpose or its programmatic goals. As a result, the organisation may simply want to get the project completed rather putting effort into delivering the best outcome possible.

2. When small organisations are composed of hobbyists. This scenario often involves small groups of devoted individuals who self-fund their own projects. Typically, the effectiveness of the project is not the primary reason for the group undertaking the project. Generally, the people involved are more interested in enjoying the process rather than focusing on the quality of the outcome. Consequently, the organisation's members may be somewhat disinterested in improving their efficiency.

3.4 Recommended resources

- https://www.conservationmeasures.org/ A portal for conservation projects which gives detailed and comprehensive advice on designing, measuring, and implementing conservation programmes.

- http://www.tools4dev.org/ A website providing many resources for building the skill set of international development and aid professionals.

- https://whatworks.org.nz/ A website which describes in simple terms different project design frameworks and approaches.

- Linda Tuhiwai Smith (2012) Decolonizing methodologies: Research and Indigenous Peoples, second edition. Zed Books, London. A book specifically for indigenous researchers working in indigenous contexts. The book's central critique can easily be extended to conservation and development projects.

3.5   Summary

- Most modern organisations use some form of adaptive management cycle to improve their performance.

- Currently, many project donors expect to see the project's success reported against project indicators (often outlined in a logframe).

- Reporting on project indicators requires the collection and analysis of data.

- Programs like R offer a pathway to produce the project indicators in an accurate, consistent, and fast manner.

- The Plan phase is the most important for data integration to ensure:

  - A budget has been allocated for the data analysis.
  - The basic analytical methods have been worked out and documented.
  - Staff have been allocated to undertake the data analysis.
  - A strategy to communicate the project's results with external decision makers has been made.

4

Getting started in R

In this chapter we will cover:

- Downloading and installing R.

- The R interface.

- Ways to write and organise code.

- How R is organised.

4.1 Installing R

Getting R installed on your computer is simple but will require a good internet connection. Alternatively, you may be able to get an electronic version from a colleague with the same operating system.

The steps are:

1. Download an R version that is appropriate for your computer's operating system (e.g. Windows, Mac, or Linux distribution) from https://cran.r-project.org/.

2. Install R. Unless you know the modifications you need leave all default settings in the installation options as they are.

    - **Linux:** choose your distribution and follow the instructions on the website.

    - **Mac:** save the .pkg file, double-click it to open, and follow the instructions.

    - **Windows:** save the .exe file, double-click it to open, and follow the instructions. You will be offered an option of selecting between a 32 bit or 64 bit versions to match your computer (you can find this information by checking the 'properties' of your computer). If you are unsure you can install both.

This is all you need to start running R. However, I would recommend you download and use a integrated development environment (IDE). An IDE makes writing code like R easier. An IDE provides us with a range of tools for writing, saving, and debugging code. For example, if you miss a bracket or a comma in a line of code an IDE will let you know that there is a mistake, and may even suggest a correction.

4.2   Installing RStudio

A good IDE for R is called RStudio. RStudio is a freely available and works on Windows, Mac, and Linux operating systems. Once you have installed R you can download and install the RStudio version that is appropriate for your operating system at `https://rstudio.com/products/rstudio/download/`.

IDE interfaces are somewhat more complicated than that of a simple script editor. To a beginner this can be a little overwhelming. But unlike the the default script editor IDEs have inbuilt debugging functions. Simple script editors can't do this.

4.3   The R interface

When we open the default version of R, it will show only one window called the console (Figure 4.1). A separate window for a script editor will appear when it is created, as will a graphics window for graphs.

By comparison when we open RStudio it will show three windows (called panes) and the console on the left hand side (Figure 4.2). Again, just like default R the script editor will only appear once it is created (appearing above the console). The two other panes show helpful information about your R session. The significance of these will be come apparent as your knowledge of R increases. The most important of these panes are:

1. **Environment** (top right): shows the objects that have been created in your session.

2. **History** (top right): shows the sequence in which your code was run.

3. **Files** (bottom right): is a way of viewing files (defaults to you project folder).

4. **Plots** (bottom right): is where any graphs created will appear.

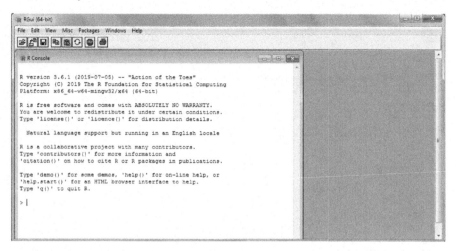

FIGURE 4.1
Screen shot of the default R interface

5. **Packages** (bottom right): shows the packages available and loaded into your session.

6. **Help** (bottom right): is where help files will appear when requested. The home button leads to a contents page which holds lot of R information for both beginners and advanced users.

4.3.1 The console

Regardless of whether we are using the default version of R or RStudio the console is where you will find basic information about the version of R you are running. Any code we try to run in a script editor will appear in the console along with any numerical or text outputs, error messages, or warnings associated with the processing of the code.

4.3.2 Version information

Once you have opened R you will see that version information about R appears in the console window. This information is important as there can be compatibility issues with older versions of R. The version is represented by 3 numbers separated by two decimal points followed by a date e.g. R version 3.6.1 (2019-07-05). Sub-minor version changes are shown by the final number – these usually will not have any impact on how your code runs. Major and sub-major version changes are shown in the first two numbers. Older versions of R may not be able to run code (especially packages) written in newer versions of R (especially if there has been a major version change).

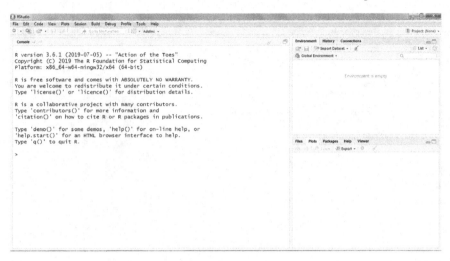

FIGURE 4.2
Screen shot of the RStudio interface

New major versions of R come out about once a year. So it is a good idea to update your version of R at least yearly. Updating R will not remove any of your saved R files.

4.3.3 Writing code in the console

We can write R code directly into the console beside the prompt '>'. We can try this now. Let's write 3 + 2 and press **ENTER**:

```
> 3 + 2
```

This will result in the following output in the console:

```
[1] 5
```

The '[1]' is an index which shows the number of elements returned in the output. In our example, the output has one result which has the value of 5. Sometimes we can have hundreds of results but rather than giving the index for each element, R only shows the index number for the element nearest the left hand margin of the console. For example, if we wanted a sequence from numbers from 1 to 50 we would use the following code:

```
> 1:50
```

And get the following output:

```
[1]   1  2  3  4  5  6  7  8  9 10 11 12 13 14 15 16 17 18
[19] 19 20 21 22 23 24 25 26 27 28 29 30 31 32 33 34 35 36
[37] 37 38 39 40 41 42 43 44 45 46 47 48 49 50
```

4.3.4 Script editors

Writing code directly into the console is not the most practical way of writing code. A better way to run R code is to use the script editor. A script editor allows us to view, save, and edit the code we are working on. We can select single lines, portions, or the entire script to be sent to the console. It is important we save the script with a descriptive name – once you have large numbers of old scripts finding the script you want can become difficult.

> **Note:** If you are running R on Linux you may find there is no automatic default script editor upon opening. As a result you probably will have to install an IDE manually.

4.3.5 Using the default script editor

In the default script editor we create a new script editor by going to the R tool bar and choosing:

- **Mac:** `File > New document`
- **Windows:** `File > New script`

The script editor will now appear as a window somewhere on our screen. We can move it by clicking, and dragging the top of the window to a new position. We can simply highlight the lines of code we are interested in and send them them to the console for processing by pressing:

- **Mac:** `Command + ENTER`
- **Windows:** `CTRL + R`

4.3.6   Using RStudio

Alternatively, rather than using R with the default script editor we could use RStudio. One of the big advantages of RStudio is that it allows us to create projects – a way of organising our R files in a folder system. Projects allow us to keep our scripts, data files, text, and any outputs in one place. I recommend you keep have a specific folder on your computer where you keep all your R code. Within this R folder I recommend you have series of RStudio project folders which correspond to the real life project you or your organisation are working on. Ensure that you give each project a meaningful name so that you can easily find it in 5 years time. Creating a project sets the project's working directory – meaning RStudio always knows where to look for the files connected to a particular R project. To create a project we follow these steps:

1. Choose `File > New Project` from the R tool bar.

2. In the pop window choose `New Directory > New Project`.

3. Create a name for your project in the `Directory name` field.

4. Press `Create Project` when finished.

RStudio then creates a folder into which you can store all the files associated with the project, and takes you to a new screen. You'll notice that the path to the project appears above the RStudio tool bar and the name of the project appears the far right of the toolbar. RStudio also has its own script editor. To get a new script editor we choose:

- `File > New File > R Script`

This editor behaves exactly the same as the default editor except to run the code we have to highlight and press:

- **Linux**: `CTRL + ENTER` or `Run` from the upper right of the script window.

- **Mac**: press `Command + ENTER` or `Run` from the upper right of the script window.

- **Windows**: `CTRL + ENTER` or `Run` from the upper right of the script window.

4.4   R as a calculator

Okay – let's start using with our script editor! R can be used as a calculator. Simply write the calculation in the script editor and run the line:

```
3+2
```

The code and the subsequent result will appear in the console window:

```
3+2

[1] 5
```

All lines are run in the order that they are received by the console. For example, if we write the following lines and run them:

```
3 + 2
6 * 10
```

Our results are given as:

```
[1] 5

[1] 60
```

Any code we run will always appear in the console along with any output, error messages, or warnings. If the code results in a graph this will appear in the plot pane of RStudio, if you are using default R it will appear in a pop-up graphics window.

4.5 How R works

To start using R you need to know about the 3 basic levels of organisation in R:

- Objects
- Functions
- Packages

4.5.1 Objects

In R we work with objects. Objects are a way of storing data in R's memory. Every object must have a name. If we are using RStudio the names of all the objects in R's memory will appear in the Environment pane.

So, let's start by making two simple objects. To do this we use the assignment operator '<-' which is the less than sign followed by a minus sign. This makes an arrow shape. We can think of the assignment operator being the same as the equals sign '='. Indeed, you can use the equals sign if you prefer (as in the example below). However, I find the assignment operator to be better as it stands out when code becomes more complex.

```
a <- 3
b = 2
```

Once those two lines are run we will have two objects called a and b. If we run the name of the object R will show us the data associated with it:

```
a

[1] 3
```

We can also do mathematical procedures with objects:

```
a + b

[1] 5
```

4.5.2 Functions

R uses functions to undertake specific tasks. There are thousands of functions in R. Perhaps the most commonly used function in R is the **c()** function. The 'c' is short for 'concatenate' which means to link in a series. We could use the **c()** function to make an object called weight made up of a and b:

```
weight <- c(a, b)
```

We can then confirm what data the object `weight` contains:

```
weight

[1] 3 2
```

As R is a statistical programming language it has lots of inbuilt statistical functions like **mean()** for finding the mean (also known as the average). So to find the mean of `weight` we use:

```
mean(weight)

[1] 2.5
```

4.5.2.1 Getting help on functions

Every function in R has something called arguments. Arguments tell us what information a function needs (within its brackets). To find out what arguments a function requires, its options, as well as more information we type a '?' followed the function we want information on. For example, if we wanted to know more about the **mean()** function we would run:

```
?mean()
```

The function will then access a document already stored on our computer. It is not accessing the information via the internet. Sometimes the help documents are not as helpful as we would like. In these situations we can usually find some helpful examples on the internet by searching the name of the function together with 'R'.

4.5.3 Packages

Functions are contained within packages. Base R contains quite a number of packages pre-loaded. To see these you can run the following code:

```
installed.packages()
```

More advanced R users can write their own functions. Often they will share these functions as packages so other people can benefit from their code. Many of these packages can be accessed directly via R. For example, to install the `tidyverse` package (which we will use heavily in later examples) we would run the following code:

```
install.packages("tidyverse")
```

R may then ask us if it is okay to install the library in a particular location – usually it is fine to accept (this will only happen the first time you download a library).

> **Note:** the `tidyverse` package is rather large and could take some time to download if you have slow download speeds.

Alternatively, you can load these packages from your R interface:

- In the default R interface click the 'Package' drop-down menu from the Tool bar and click 'Install package(s)'. You will also be asked to choose a CRAN (Comprehensive R Archive Network) mirror site to download from. Typically, you want to choose a location close to us. Alternatively we can choose the '0-Cloud' option (if available) for R to make the selection for us.

- In RStudio you go to the bottom right pane and click on the 'Packages' tab. Then click 'Install' and write the name of the package you want to install. RStudio will determine the CRAN mirror site automatically.

Once downloaded the package will be stored on your computer. However, it will only be loaded in your session when you run the **library()** function. To load the `tidyverse` package we would run the following code (note: the name of the library works both with and without the use of quotation marks):

```
library("tidyverse")
```

If you upgrade to a new version of R, you may find that packages no longer work. For major and sub-major versions, a new library folder will have been made – R will be looking in this folder for your packages, however, as it will be empty R will return an error. One way around this is to manually copy and paste the packages from the old library folder (automatically named by the version number e.g. 3.5) into the new folder (e.g. 3.6). If compatibility issues remain you will likely have to re-install the latest version of packages.

4.5.3.1 Getting help on packages

Usually we won't know the names of the functions in a package nor how to use them. One way to find out this information is to go to the CRAN website which relates to the package e.g.: `https://cran.r-project.org/web/packages/tidyverse/`. You can safely ignore most of the information here – you just need to find the reference manual (as a downloadable pdf) listed under 'Downloads'. The reference manual will contain a description of all the package's functions. Many packages also have their own dedicated website which are normally listed in the CRAN information (under 'URL'). A search of the package name on the internet will usually reveal helpful tutorials which are often easier to understand than the reference manual.

4.6 Writing meaningful code

If you are like me then lines of code are not particularly memorable. Our challenge is to make our scripts understandable to a forgetful, confused future version ourselves. With this mind here are some tips to making your code a bit more helpful to future you:

- Give your script a meaningful name that will help you understand why you wrote it first place.

- Use '#'s to describe the different steps you are taking and what the code is doing. Anything written after a '#' will not be read by R. They can even be used on a line of code without any ill-effect:

```
# Anything I write here will not be read by R
# We can use any number of lines if we want
#
mean(weight)  # It doesn't affect the R code at all:
[1] 2.5
```

- Use line spacing to group and separate lines of code. Group code with similar purposes together and separate these from other lines of code by line breaks.

- Make your object names lower case. R is case sensitive and it is easy to make mistakes with capitalised object names (see Chapter 13).

- Avoid giving objects the names of existing functions – this will create unnecessary confusion.

- Put spaces around commas and operators like <-, =, + etc. to make them readable, just like we would with normal punctuation.

- Use '<-' rather than '=' to assign names to objects as the '=' sign is often used inside functions.

- Consider writing small code examples which explain how to do a task in R which you are sure you will forget. Save these as 'How to ***.R' files where *** is the name of the task. This is an easy way of helping future you – this way you can just search your 'How to' R files for the name of the task. Hunting through hundreds of your own files for a small code snippet you used several years ago is no fun.

4.7   Reproducibility

An ongoing issue within conservation and development projects is the loss of project and institutional knowledge. The R code that we produce to analyse the effectiveness of our projects is an important part of that knowledge. Any loss of project or institutional knowledge affects the ability of organisations to reproduce good outcomes and avoid bad ones. The loss of knowledge comes from:

1. Staff turnover.

2. Project methods not being documented.

3. Projects not being evaluated.

4. Data not being safeguarded.

5. Analyses not being safeguarded (i.e. our R code).

6. Leadership not investing in a long term data strategy.

Multi-year projects and programmes often outlast many of the people working on them. In addition, in most conservation and development organisations, different people are used to collect, enter, and analyse the data. What might be an obvious way of collecting or recording data for one person might be completely confusing to another. A spreadsheet, by itself, is not interpretable without some kind of context. The absence of a written method (describing how the data was collected and how variables were recorded) may prevent some projects from being evaluated properly while poor data collection may mean no evaluation is possible at all.

Many conservation and development organisations work in countries with security issues. Burglary of offices, theft of personal laptops, hard drives, and bags is a common occurrence. For this reason alone, data (regardless of format), needs to be backed up off-site. Another issue related to data security are problems caused by having so many versions of the data circulating that there is no master version, or worse, a number of disputed master versions. Ideally, to overcome these issues we would want to safeguard our data in a single secure database. The reality is, however, that such systems are not within the budget of most smaller organisations. However, there are a number of free or low cost tools which can help to overcome the same issues. For example, Kobotoolbox (`https://www.kobotoolbox.org/`) is a free data collection tool developed specifically for researchers and humanitarian organisations working in developing countries. Kobotoolbox and its apps allow data to be collected by smartphone and uploaded to the internet in near real time. Version control of other data can be simplified by using common online file sharing platforms. Not only can these platforms backup data automatically, but many can ensure that the most recent version is the default version for organisational sharing (while also ensuring earlier versions can be accessed if needed).

In a similar way, analyses also should be safeguarded. Not only does this mean we have a record of our analyses, but as organisations tend to reuse the same kinds of monitoring and evaluation methods it means our analyses tend to be the same. So it makes sense to keep our R code so that we can copy and change existing code rather than writing new code every time. For this reason we should name R files using searchable terms and describe our R code using '#' tags. This way other people (and our future selves) can come back to code years later and understand it[1]. Beyond this, RStudio makes storing our analyses easy through something they call 'projects'. A project is created by choosing `File > New Project` from the RStudio toolbar and following the prompts (see section 4.3.6). The 'project' then acts like a folder where we can store our R scripts, data, and any related information (like written methods) relating to that particular project. One of the great advantages of RStudio projects is we do not have to tell R where to find the files we are working with (this is known as the working directory – an alternative approach is to use the **setwd()** function, see Chapter 5 for more detail). This means that if we share the project folder with anyone they will be able to run the code without having to make any changes to the code.

The last item on our list, investing in a data strategy, is likely the hardest. A data strategy is a knowledge strategy – a way of safeguarding organisational knowledge beyond individuals so that can be stored, accessed, analysed, shared, and used to inform organisational decision making. However, managing the day-to-day struggles of a conservation or development organisation is hard work, so developing a long-term data strategy is never an immediate priority. Unfortunately, time eventually catches up, and even information that might seem common knowledge today will eventually be lost. Consider this simple test: imagine your organisation in 10 years time: all your old staff have moved on and now your organisation wants to start working with a 'new community'. Ask yourself this: 'How would the new staff know if your organisation

---

[1]if you want to go one step further you could try RStudio's 'R Notebooks' as a way of combining code and text as a way of making semi-automated reports.

had or hadn't worked in this community in the past?' Regardless of whether or not your organisation remembers, that community would certainly not have forgotten.

4.8   Recommended resources

- `https://rstudio.com/resources/cheatsheets/`. This website by RStudio provides a number of short downloadable guides ('cheatsheets') including a handy 2 page pdf guide on the RStudio IDE.

- John Verzani (2011) Getting started with RStudio. O'Reilly Media, Sebastopol.

- Alastair Stark (2020) Institutional Amnesia and Humanitarian Disaster Management Working Paper 005. An overview of the loss of institutional knowledge and how it affects humanitarian agencies.

4.9   Summary

- Run your code from the script editor rather than the console.

- Consider using an IDE (e.g RStudio) to make coding easier.

- In order to start using R you need to know the differences between objects, functions, and packages.

- Think about your R code as part of bigger knowledge system. Work with others to make to make that system successful.

5

Introduction to data frames

In this chapter we will explore:

- Making a data frame.

- Importing a data frame.

- Saving a data frame.

- Making a reproducible example.

5.1 Making data frames

The data frame is the fundamental unit for doing data analysis in R. A data frame is kind of like a spreadsheet. It is a table-like form of data which R can read (shown in Figure 5.1). If we are developing a spreadsheet for people we tend to design it so that it is easy to read for people. This differs to how we design a spreadsheet for a computer to read. Computers handle data best when it is in a tidy form.

The rules to tidy data are simple:

- Every column is a variable.

- Every row is an observation.

- Each cell consists of a single value.

On of the easiest ways to understand a data frame is to learn to make one. This way you can see how the basic components of a data frame fit together. Let's start by making an object called **person** composed of the names of 7 people: Sione, Michiko, Marama, Oka, Talia, Zari, and Trevor.

We will use the **c()** function – the 'c' is short for 'concatenate' which means to link in a series. The **c()** is probably the most common function you'll come across in R. As these names are text (not objects) we have to put these in quotation marks.

	Variable 1	Variable 2	Variable 3
Observation 1	value	value	value
Observation 2	value	value	value
Observation 3	value	value	value
Observation 4	value	value	value

FIGURE 5.1
The basic layout of a tidy data frame

```
person <- c("Sione", "Michiko", "Marama", "Oka", "Talia",
            "Zari","Trevor")
```

The '<-' is known as an assignment operator. This is just an R way of saying 'is'. On the left hand side of the assignment function is the object name, and on the right hand side is how that object's value was calculated. If we just write `person` and run the code it will return the values associated with the object. This is the simplest type of data structure in R which is called an 'atomic vector' or just a 'vector' for short. A vector is simply a sequence of data elements:

```
person

[1] "Sione"    "Michiko" "Marama"   "Oka"      "Talia"
[6] "Zari"     "Trevor"
```

If you imagine we rotate our vector from being horizontal to being vertical – what we have done is build the first column of a data frame (Figure 5.2). So far so good. Now, let's start building a data frame by adding some more variables. Let's add the person's score in a test, and whether or not the person has been trained. As these entries are numbers we don't use quotation marks.

	person
Observation 1	Sione
Observation 2	Michiko
Observation 3	Marama
Observation 4	Oka
Observation 5	Talia
Observation 6	Zuri
Observation 7	Trevor

FIGURE 5.2
A column in a data frame is a vector

Let's name the variables `score` and `trained` and use a 0 or 1 to represent whether or not the person has been trained:

```
score <- c(165, 153, 183, 147, 177, 161, 172)
trained <- c(0,1,0,1, 0,1,1)
```

Now we can make our first data frame which we can call `my.df` by using the **data.frame()** function:

```
my.df <- data.frame(person, score, trained)
```

This will make an object called `my.df`. Data frames are always rectangular in shape. That means that each vector which makes up the data frame must be of the same length. If the vectors of different lengths are used the **data.frame()** function will return an error. If we want to see the entire data frame (or any object) we just highlight its name and then run it (however, if the data frame is very large we definitely don't want to do this).

5.2   Importing a data frame

While it is good to understand how to simulate a data frame (e.g. for purposes of making a reproducible example, understanding R, and practicing modelling approaches) most of the time we will be wanting to work on our real data. To do this we need to know how to import data. The easiest way to get data in R is to follow these steps (however, if you are using a 'project' in RStudio you can skip steps 3 and 4 by simply putting the file into your project folder):

1. **Step 1:** Build your data in a spreadsheet using the principles of tidy data (i.e. every column is a variable, every row is an observation, each cell consists of a single value)

2. **Step 2:** Save your spreadsheet as as '.csv' file (typically found in the 'save as' option of your spreadsheet program). This file type is short for 'comma-separated values'.

3. **Step 3:** Copy the path (location) of your csv file (visible in the properties or information associated with your file).

4. **Step 4:** Paste the path of the file into the **setwd()** function. Remember to include quotation marks. The function is short for set working directory. The configuration of the **setwd()** function differs according to your operating system:

   Windows: Copying and pasting the path into an R script results in just a single backslash. For this to be read by R, an extra backslash has be inserted. Alternatively, a single forward slash can be used:

   ```
   setwd("C:\\Users\\Me\\Documents")
   # or
   setwd("C:/Users/Me/Documents")
   ```

   Mac: Copying and pasting the path into an R script results in just a forward backslash - so no change is necessary.

   ```
   setwd("/Users/Me/Documents")
   ```

   Linux: Copying and pasting the path into an R script results in just a forward backslash – no change is necessary.

   ```
   setwd("/usr/me/documents")
   ```

5. **Step 5:** Paste your file's name into the **read.csv()** function to read your file. Don't forget to put '.csv' at the end of your file's name.

```
my.df <- read.csv("My spreadsheet data.csv")
```

Note: In countries where the decimal place is a period **read.csv()** should be used. In countries where the decimal place is a comma **read.csv2()** should be used instead. Additionally, despite being called 'comma-separated values' in the later situation a semi-colon rather than a comma is used to separate columns.

5.3   Saving a data frame

Saving a data frame as a '.csv' file is straight forward. You just need to give the **write.csv()** function the name of the object, and the name you want to give to the file you are saving. Most of the time you will want to use *row.names = FALSE* argument to avoid a column of reference numbers being added. By doing this your saved file will look exactly like the one you made in R:

```
write.csv(my.df, "My R data.csv", row.names = FALSE)
```

This will save the file My R data.csv to the working directory (in the path described by **setwd()**). You can always check the path by using the **getwd()** function (get working directory):

```
getwd()
```

5.4   Investigating a data frame

We can use the **class()** function to confirm my.df is a data frame. If we wanted to see what the class of a variable within the data frame is (such as **trained**) we use the '$' operator after the name of the object to reference a particular variable:

```
class(my.df)

[1] "data.frame"

class(my.df$trained)

[1] "numeric"
```

So we see that R will treat `my.df$trained` as a numeric variable. Now in this particular case perhaps we had imagined that is actually just a shorthand for two categories that exists as '0' (no), or '1' (yes). Categorical data is usually handled best in R by defining them as a factors. We can do this by using the **as.factor**() function and telling R the name of the data set and variable we are wanting it to reassign:

```
my.df$trained <- as.factor(my.df$trained)
```

Now to save time rather than use the **class**() function we can just use the **str**() function to get an overview of the structure of the object to see how the various data components are being treated by R:

```
str(my.df)
```

```
'data.frame': 7 obs. of 3 variables:
$ person : Factor w/ 7 levels "Marama","Michiko",..: 4 2 1 3 5
7 6
$ score : num 165 153 183 147 177 161 172
$ trained: Factor w/ 2 levels "0","1": 1 2 1 2 1 2 2
```

The output tells us that `my.df` is a data frame made up of three named variables: `person`, `score`, and `trained`, which will be handled by R as a factor, a numeric, and factor. The output also provides some details of each variable, the number of levels (categories) of a factor, and a sample of some of their initial values.

5.5 Other functions to examine an R object

There are a few core functions that are very useful for examining objects. Sometimes objects will be very large data frames made up of hundreds, thousands, or millions observations. In such cases we will not want to view the whole object in the console.

The function **head()** provides an overview of the first 6 lines of the data frame. This is particularly helpful if we want to quickly check the overall look of a data frame. It also helps us confirm that we have created our data frame in the way we had hoped.

```
head(my.df)

    person score trained
1    Sione   165       0
2  Michiko   153       1
3   Marama   183       0
4      Oka   147       1
5    Talia   177       0
6     Zari   161       1
```

Similarly, we may want just to see the last 6 lines of the data frame. In this case, the opposite of **head()** is **tail()**.

```
tail(my.df)
```

Sometimes we may want just to see the unique column names using the **names()** function. This is particularly helpful for very large data frames where functions like **str()** or **head()** may produce just too much information.

```
names(my.df)

[1] "person"  "score"    "trained"
```

The **dim()** function tells us the dimensions of an R object. For a data frame this represents the number of rows and columns. This is usually a good place to start if anyone is asking you to help them with their data set. Often people will give you the wrong data, and one of the easiest ways to ensure it is not an incorrect version is to get them to confirm the number of rows and columns expected. The output below confirms our data frame has 7 rows and 3 columns of data (in R rows always come before columns).

```
dim(my.df)

[1] 7 3
```

R is case and punctuation sensitive. Consequently, small errors such as a rogue space or inconsistent capitalisation during data entry will result in R reading a different number of categories than expected (called levels in R). Often we will want to check that the number of categories of a variable matches our expectations though the **unique()** function. For example, we can check that there are only two levels in `my.df$trained`:

```
unique(my.df$trained)

[1] 0 1
Levels: 0 1
```

5.6   Subsetting using the '[' and ']' operators

In R values in a data frame are referenced by their location in terms of rows and columns using the '[' and ']' operators. Taking the form:

```
my.df[row, column]
```

This then allows us the ability to manually subset data sets[1]. For example, the 3rd row of the 1st column would be:

```
my.df[3, 1]
```

---

[1]to make the subsetted data permanent we would have to overwrite our object e.g. `my.df <- my.df[3, 1]`

```
[1] Marama
Levels: Marama Michiko Oka Sione Talia Trevor Zari
```

We can select a row of the data frame by giving the row position but leaving the leaving the column position empty:

```
my.df[3, ]

  person score trained
3 Marama   183       0
```

We can select a column of the data frame by giving the column position but leaving the leaving the row position empty:

```
my.df[ , 1]

[1] Sione   Michiko Marama  Oka     Talia   Zari    Trevor
Levels: Marama Michiko Oka Sione Talia Trevor Zari
```

Values can be removed by using a negative sign. For example, we can remove the 1st row of our data frame:

```
my.df[-1, ]

   person score trained
2 Michiko   153       1
3  Marama   183       0
4     Oka   147       1
5   Talia   177       0
6    Zari   161       1
7  Trevor   172       1
```

Multiple rows or columns can be removed or included by using the **c()** function. By including a '-' sign before the **c()** function we can remove a number of rows:

```
my.df[-c(1,2,5,6,7), ]

   person score trained
3  Marama   183       0
4     Oka   147       1
```

The same principle extends to vectors. Except of course there is only one dimension (not two). So to see the first three entries of `my.df$person`:

```
my.df$person[1:3]

[1] Sione   Michiko Marama
Levels: Marama Michiko Oka Sione Talia Trevor Zari
```

We can also use '[' ']' to subset some built-in constants in R such as: **letters**, **LETTERS**, **month.name**, and **month.abb**:

```
LETTERS[1:10]

[1] "A" "B" "C" "D" "E" "F" "G" "H" "I" "J"

letters[1:10]

[1] "a" "b" "c" "d" "e" "f" "g" "h" "i" "j"

month.name[1:5]

[1] "January"  "February" "March"    "April"    "May"

month.abb[1:8]

[1] "Jan" "Feb" "Mar" "Apr" "May" "Jun" "Jul" "Aug"
```

5.7 Descriptive statistics

We can easily get some descriptive statistics for a variable (e.g `my.df$score`) by using some simply named statistical functions: the mean with **mean()**, the median with **median()**, the range with **range()**, and the standard deviation with **sd()**. For example:

```
mean(my.df$score)

[1] 165.4286
```

Or if we want, we can be lazy and use the **summary()** function to get a summary for every variable in the data frame. This function gives basic descriptive statistics for the numeric variables, and counts for the factor variables:

```
summary(my.df)

    person        score       trained
 Marama :1   Min.   :147.0   0:3
 Michiko:1   1st Qu.:157.0   1:4
 Oka    :1   Median :165.0
 Sione  :1   Mean   :165.4
 Talia  :1   3rd Qu.:174.5
 Trevor :1   Max.   :183.0
 Zari   :1
```

5.8   Viewing data frames

A data frame can always be viewed in its entirely in the R console. However, it starts to become impractical when the data frame is larger than the console window. Often, it is helpful to be able to view the whole data frame outside of the console. There are a couple of ways to do this:

- In base R the data frame is viewable as a table in a separate window using the **fix()** function (the fix window must be closed otherwise it will prevent code from running). Note: this function will not run in RStudio:

```
fix(my.df)
```

- In RStudio the data frame can be viewed as a table in the editor pane: using the **View()** function:

```
View(my.df)
```

- Alternately, in RStudio we can simply click on the entry which corresponds to the data frame in the Global Environment pane and the data will be shown in a new tab in the editor pane.

5.9   Making a reproducible example

If you are a frequent R user at some point you will come across a coding issue specific to your data or analysis which isn't solved by any of the reference material you have. In this case you will have to reach out for help. Whether it is emailing a colleague or posting a query to a website, the easiest way to get the right kind of help is to make a reproducible example.

Importantly, a reproducible example is a tool for others to help you learn how to solve a specific issue, not to get other people to do your analysis for you. As the name suggests, the example also needs to be reproducible by others so that they can understand the type of issue you are facing.

`Stackoverflow.com` is a popular site which specialises in helping people solve coding problems. They have compiled a basic list of what makes a good reproducible example (which I have lightly edited below):

1. A minimal data set which replicates the data types involved.

2. A short piece of runnable code which shows the issue.

3. A seed for anything involving random number generation (this will allow the same results to be reproduced by other people).

4. In the case of suspected compatibility issues: information on the packages used, R version, and the computer's operating system.

Often, just by making a reproducible example, we will end up solving the issue ourselves.

5.9.1   Reproducible example steps

If we wanted to make a reproducible example that replicates an issue about changing a numeric variable to a factor we would do the following:

- **Step 1:** Give the issue a title so that other people who can help, or are experiencing a similar problem can find it e.g. 'Changing a numeric variable to a factor in R'.

- **Step 2:** Outline the nature of the issue so that the people trying to help you know what you are trying to do: 'If I have a column in a data frame being read as numeric how do I turn it into a factor?'

- **Step 3:** Give the code for the minimal data set. Rather than making a simulated data frame in full like we did for `my.df` we can make a minimal data set using some of the functions for generating samples, numbers, and letters.

As we want a minimal data set we don't need the full 7 rows of data. Rather we could just create a data frame with 3 rows of observations. Also we don't need the `score` variable so we can leave it out entirely. Rather than making a full set of names, we can just use a capital letter instead (by using the **LETTERS** constant). For the `trained` variable we can just use R to randomly sample from two options (0 or 1) with replacement, three times using the **sample()** function. Generating random data can create a problem with reproducibility. For this reason we should use the **set.seed()** function. By providing a starting number (known as a seed) to this function anyone using our code will be able to get exactly the same set of randomly generated numbers.

Setting the random seed:

```
set.seed(83)
```

Creating 2 variables:

```
name <- LETTERS[1:3]
trained <- sample(c(0,1), 3, replace = TRUE)
```

Turning the variables into a data frame:

```
mydata <- data.frame(name, trained)
```

Show the issue R (e.g. reads column as numeric not as a factor):

```
# R reads column as numeric (not as a factor)
class(mydata$trained)
```

```
[1] "numeric"
```

- **Step 4:** If you think the issue relates to a comparability issue you should give the output generated by the **sessionInfo()** function.

```
sessionInfo()
```

```
R version 3.6.1 (2019-07-05)
Platform: x86_64-w64-mingw32/x64 (64-bit)
Running under: Windows 7 x64 (build 7601) Service Pack 1

Matrix products: default

locale:
[1] LC_COLLATE=English_Australia.1252
[2] LC_CTYPE=English_Australia.1252
[3] LC_MONETARY=English_Australia.1252
[4] LC_NUMERIC=C
[5] LC_TIME=English_Australia.1252

attached base packages:
[1] stats     graphics  grDevices utils     datasets  methods
[7] base

loaded via a namespace (and not attached):
[1] compiler_3.6.1 tools_3.6.1
```

5.10   Recommended resources

- https://cran.r-project.org/doc/manuals/r-release/R-intro.html. The official introduction manual for R.

- https://stackoverflow.com/questions/5963269/how-to-make-a-great-r-reproducible-example. A guide from the Stackoverflow community for making a minimal reproducible example.

5.11 Summary

- The simplest unit of data in R is a vector.

- Data frames are made up of equal length vectors.

- The best way to understand data frames is to make them.

- We need to be familiar with the '$' and '[' and ']' operators so that we may reference specific data elements.

- We need to be able to make data frames in order to make reproducible examples.

6

The Waihi project

Throughout this book we will use a fictitious example of a remote island somewhere in the Western Pacific between the countries of Papua New Guinea and the Federated States of Micronesia. This is the island of 'Waihi' where a conservation and development project is taking place. Every example is fictitious, neither the place, the people, nor the projects exist. In this chapter we cover:

- The Waihi project scenario.

- Installing the `condev` package.

- Using the data examples used in this book.

6.1 The scenario

Our imaginary project on Waihi is constructed as a typical community conservation project. Like many remote communities existing today, the imaginary inhabitants of Waihi would like to access the benefits of the modern world such as basic health care and an opportunity to earn a cash income. However, this imagined forest community has only one major resource with which to earn a cash income – timber from old growth forests. The communities of Waihi are well aware of the value of the forest to their subsistence livelihoods as it provides: timber for their houses and canoes, medicinal plants for their ailments, habitat for their game animals, as well as areas of cultural importance.

In this scenario, some concerned community leaders have approached our organisation for assistance in both safeguarding their forest resources and developing a sustainable cash income. A process of free, prior, and informed consent follows. Subsequently, an agreement is reached and the terms of engagement agreed upon between the organisation and the community. In the next step our organisation undertakes participatory planning with the community. Through this process the project's objectives are decided upon and a set activities designed to achieve them. Based on this information a logframe is developed and a proposal submitted, and then, after some follow up discussions with a donor – funded.

6.1.1   Why evidence is important

Skepticism, ideological differences, and conflicts of interest are features of all human communities. Village communities, like those of Waihi, are no different. Not all community members will be supportive of a project. Some community members will be deeply suspicious of our organisation and its motives. Therefore, it is important that our organisation operates in a participatory manner and communicates the results back to the community quickly, honestly, and in a manner which they can understand. While quantitative monitoring associated with a project may only occur periodically, we can be sure that qualitative monitoring of the project, by the island's communities, is happening continuously.

Our challenge, as practitioners, is to deliver the best project possible. Not only does that mean ensuring our project delivers its promised outcomes, but it also means engaging with the community and the donor in a manner which has mutually agreed upon.

Ultimately, the success of a community project is as much about the conduct of the organisation with the community as it is about achieving the targets outlined in the logframe.

6.2   The data

All of the data regarding the 'Waihi' project is located on a package called condev which is held in GitHub (which is an online repository for code). The condev package can be downloaded via code directly in R using the following steps:

- **Step 1:** First install the R package remotes from CRAN (you only need to do this once) using the **install_packages()** function:

```
install.packages("remotes")
```

- **Step 2:** Then install the condev package using the **install_github()** function. These two lines of code only need to be run once to install it on your computer:

```
library("remotes")
install_github("NathanWhitmore/condev")
```

- **Step 3:** Either hash out the installation codes lines or delete them once the installation has completed. Then load the newly installed `condev` package which contains the data for the examples using the **library()** function:

```
library("condev")
```

- **Step 4:** Check that the data has been loaded. Any data sets available for use, including in built data sets, are viewable by running **data()** function without any arguments. To view the data sets connected with a particular package we use the *package* argument:

```
data(package = "condev")
```

Which results in the following list:

```
attendees        Training course attendance (simulated)
clinic           Patient visits to health clinic (simulated)
demo1            Small generic data frame
demo2            Larger generic data frame
fishponds        Before-After-Control-Impact study (simulated)
hunt             Hunting data (simulated)
location         Location of project communities (simulated)
messy            Messy data frame (simulated)
prosecutions     Village prosecution data (simulated)
questionnaire    Project evaluation (simulated)
split.data       Variables split across multiple
                                 columns (simulated)
traps            Monthly mongoose trap captures (simulated)
treecover        Emulation of Hansen et al. (2013)
                                 tree canopy cover data
waihi            Spatial data for the fictitious island
                                 of Waihi
```

To load a specific data set we supply the name to the **data()** function. At this point the data set becomes available for the R session. For example, to load the `demo1` we use:

```
data(demo1)
```

6.2.1   Description of `condev` data sets

attendees: a data set that simulates attendance data collected by conservation and development projects when undertaking training projects. Summaries of this type of data are usually required by donors as mandatory proof that training actually took place.

clinic: an example of a data set containing sensitive information (visits to a health clinic by pregnant women). The purpose of this data is to show we can use R to anonymise sensitive data.

demo1: a small data set to demonstrate different graphing and modelling approaches.

demo2: a larger data set to demonstrate different graphing approaches.

fishponds: a simulation of data from the Before-After-Control-Impact study design outlined in Chapter 2 (involving the testing of different kinds of fish feed). These types of study are often found in livelihood projects.

hunt: a simulation of a small hunting study. Such studies are common in countries where communities are reliant on wildlife for everyday food. The analysis of this kind of data is important for understanding hunter impacts on local wildlife and the importance of wild caught meat in local diets.

location: a small data set to demonstrate how data frames can be used to store point locations for mapping.

messy: most of the time, in conservation and development projects, the person who analyses the data is not the same person who collects the data. As a result, the person analysing the data often has to correct a variety of mistakes. This data set simulates the most common mistakes.

prosecutions: this data set simulates a simple Before-After design to determine the effectiveness of an environmental awareness program. In this example, we want know if the training of village magistrates resulted in an increase in monthly prosecutions.

questionnaire: this data set represents a simulated evaluation of a conservation and development project. This data will be used to demonstrate the usefulness of clustering and classification techniques.

split.data: this data sets will be used to demonstrates how R can be used to fix a situation in which variables are duplicated across multiple columns. Such situations can arise when multiple people are involved in data entry but use different column naming systems.

traps: this data set emulates a predator trapping program. Such programs are common across the Pacific. In this data set we are interested in knowing whether forest cover and different types of bait and trap type affect the catch rates of an invasive mongoose.

treecover: is a data set which mimics the tree canopy cover for year 2000 downloadable from the Global Forest Change website[1]. The purpose of this data is to introduce users to basic raster manipulation. It requires the installation of the `raster` package to be read – otherwise an error will result.

waihi: this data set maps the fictitious island of Waihi. The purpose of the data set is to demonstrate how different mapping techniques can be undertaken in R.

6.3 Recommended resources

- `https://github.com/NathanWhitmore/condev/`. The GitHub site for the `condev` package. Download instructions in R are given towards the bottom of the webpage.

6.4 Summary

- In order to run the examples in this book we need to install the `condev` package on your computer from GitHub (not CRAN).

- All data examples used in this book are fictitious.

- There's more to project success than just project outcomes. Every organisation is judged by their conduct.

---

[1]`https://earthenginepartners.appspot.com/science-2013-global-forest`

Part II

First steps

7

ggplot2: graphing with the tidyverse

In this chapter we will introduce:

- The `tidyverse` package.

- How to make different kinds of graph using the `tidyverse`.

- How to save a graph in the `tidyverse`.

7.1 Why graph?

Regardless of what field we work in – whether it is conservation, development, physics or finance, the starting point for understanding a data set is usually to graph it. This makes sense because:

- Most of us explore the world with our eyes.

- Graphs help us identify mistakes and patterns quickly.

- Graphing data is relatively fast.

- A good graph is an easy way to communicate our results with others.

Whether we are graphing to develop ideas or test hypotheses, a good graph tells a story. These are the stories of variables, their inter-relationships and influence (or lack of influence) on the world around us. Descriptive statistics and model equations (which we will cover in future chapters), while important, often require technical skills to be understood. On the other hand, graphs can be understood by most people with only a sentence or two of explanation. For this reason, good graphs are critical for communicating the results of conservation and development projects. As graphs can be so easily understood they are also the best starting points for coding in R – because within a few minutes you have the reward of a professional looking graph.

7.2   The `tidyverse` package

One of the most useful packages currently available for graphing is the **tidy-verse**. The **tidyverse** is a suite of interconnecting packages which allows users to produce high quality graphics, and undertake data wrangling (Chapter 9) easily. All of these packages can be installed using:

```
install.packages("tidyverse")
```

Just remember you only have to do this once. Once it is installed we can load it into our R session whenever we need it using:

```
library("tidyverse")
```

7.3   The data

For this chapter we will use three example data sets (**demo1**, **demo2** and **waihi**) from the **condev** package. Start by loading **condev** and the data sets:

```
library("condev")
data(demo1)
data(demo2)
data(waihi)
```

Note: condev is not a CRAN package and will have to be installed from GitHub – see Chapter 6 for installation instructions and descriptions of the data sets.

Whenever we import a data set our first step should always be to examine the structure of the data set using the **str()** function. This function gives us information on the data structure – importantly how R is reading the different columns. For example:

```
str(demo1)

'data.frame': 29 obs. of 5 variables:
$ x : num 10.14 9.96 10.33 9.57 10.34 ...
$ y : num 10.25 9.63 10.4 9.53 9.87 ...
$ group1: Factor w/ 3 levels "A","B","C": 3 3 3 3 3 3 3 3 3 3
...
$ group2: Factor w/ 2 levels "treat1","treat2": 1 1 1 2 2 2 1
1 2 2 ...
$ error : num -0.1402 0.3483 -0.3589 0.0115 -0.127 ...
```

What you we see is that x, y, group1, group2, and error are actual column names in the demo1 data frame. We see that two of the columns (group1 and group2) contain factors, while the rest are numeric (i.e. they contain numbers). It is important to remember that these column names are entirely arbitrary (they could have been given any name).

7.4 Graphing in R

While base R (the version of R we get without loading extra packages) can make graphs fast, it is very time consuming to make these graphs look good. Within the tidyverse is a package called ggplot2. With ggplot2, it is very easy for us to make professional looking graphs quickly. This is because ggplot2 uses a clever code structure to describe how to build a graph (the 'gg' in ggplot2 is short for the 'grammar of graphics'). Thanks to this code it is very easy to build a great range of graphs.

7.4.1 Making a ggplot

Whenever we load the tidyverse, the package ggplot2 automatically gets loaded (so we don't have to load the package separately). The package ggplot2 allows us to make a ggplot – which is a graph. A ggplot can be thought of a series of code layers (see Figure 7.1). There are slightly different ways to make a ggplot. Throughout this book we will use a form in which the data set and variables are given in each layer. I have found that this form helps prevent people getting confused about what they are trying to graph. A description of the layers in Figure 7.1 are given below:

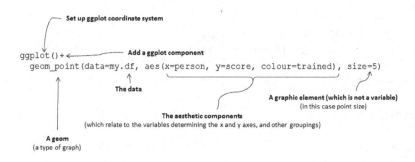

FIGURE 7.1
The basic parts of a ggplot

- **ggplot()**: sets up the coordinate system.

- **'+'**: adds a new layer.

- **geom_point()** : adds a graph type of **geom_point** (different types of graphs are made using different types of geom).

  - **data**: tells the function what our data frame is called.

  - **aes()**: these are aesthetics – the elements of the graph which relate to our variables (usually columns in our data frame).

  - outside and following the **aes()** function are any elements of the graph which we want to set for all data points.

In this way there are a vast range of graph types that can be made. In the following section we will see how they are all based on the same basic code pattern with just minor variations.

> **Note:** The graphs in this chapter have been stylised for publication and will differ slightly in look from that given by the code.

7.4.2   Scatter plots

Scatter plots are perhaps the most simplest type of plot. Each data point is simply as shown as by their x and y values. The following are 3 of the commonest variations on the scatter plot in ggplot:

- In **geom_point()** the data points are a type of shape (defaults to a circle):

```
ggplot()+
  geom_point(data = demo1, aes(x = x, y = y))
```

- In **geom_text()** the data points are text labels from a column in the data frame:

```
ggplot()+
  geom_text(data = demo1, aes(x = x, y = y, label = group1))
```

- In **geom_line()** a line connects the points in the order that they appear on the x-axis:

```
ggplot()+
  geom_line(data = demo1, aes(x = x, y = y))
```

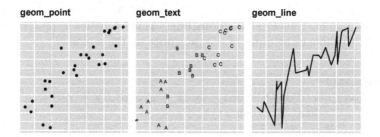

7.4.3 Bar plots

Bar plots are perhaps the most familiar type of plot. They are generally used to show totals, percentages, and proportions when comparing categories. Importantly, there are two major variations:

- In the default **geom_bar()** there is **no** *y* aesthetic – the function just creates a graph based on number of the categories in the variable given to the *x* aesthetic (with the y-axis automatically showing the total number of observations for each group):

```
ggplot()+
  geom_bar(data = demo1, aes(x = group1))
```

geom_bar

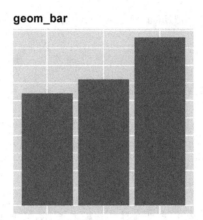

- In **geom_bar()** if we want to show the total value (rather than total number of observations) associated with each group we need to use a data frame which has a value for each instance of the group variable – this will be our *y* aesthetic. Additionally, we need to include *stat = "identity"* as an argument. This form of **geom_bar()** is much easier to deal with if we want to change the order of the categories on the x-axis:

```
ggplot()+
  geom_bar(data = demo1, aes(x = group1, y = y),
         stat = "identity")
```

In the example above, we have separated the data into the different categories found in a single variable (**group1**). If we wanted to, we could separate the data further through the use of a second categorical variable (**group2**). To do this we:

- Set the new variable (**group2**) as a new aesthetic element e.g. using the *fill* argument. This will result in stacked bar graph. In this stacked bar graph **group1** is separated across the x-axis by category, while the categories in **group2** are distinguished by their fill colour:

```
ggplot()+
  geom_bar(data = demo1, aes(x = group1, fill = group2))
```

- Alternatively, rather than having the different categories of the second variable (**group2**) stacked on top of each other we may prefer to show them side-by-side. To do this we use the argument *position = "dodge"* outside of the aesthetics:

```
ggplot()+
  geom_bar(data = demo1, aes(x = group1, fill = group2),
           position = "dodge")
```

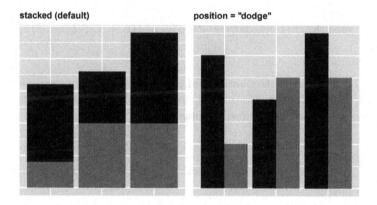

You will notice that the stacked and dodged versions emphasise very different data stories although both are showing exactly the same data.

7.4.4 Histograms

Histograms are a special class of bar plot. A histogram is made by sorting the data into categories based on the value of the variable. In a histogram the y-axis value is a measure of the number of times the category appears (i.e. its frequency). The histogram only ever takes an x aesthetic. Don't be fooled by the variable name x used in **demo1** – any continuous variable can be used. Histograms are particularly useful for checking whether a variable is skewed, and visualising differences in the distributions of different categories.

- In **geom_histogram()** the default function takes the form:

```
ggplot()+
  geom_histogram(data = demo1, aes(x = x))
```

- Alternatively, we could use **geom_freqpoly()**, which is a line version of the histogram rather than a bar plot:

```
ggplot()+
  geom_freqpoly(data = demo1, aes(x = x))
```

- In both **geom_histogram()** and **geom_freqpoly()** we can alter the number of divisions using the *bins* argument:

```
ggplot()+
  geom_histogram(data = demo1, aes(x = x), bins =15)
```

- In both **geom_histogram()** and **geom_freqpoly()** we can alter the numerical width of divisions using the *binwidth* argument (the graphs below were generated by the use of *binwidth = 1*):

```
ggplot()+
  geom_histogram(data = demo1, aes(x = x), binwidth = 1)
```

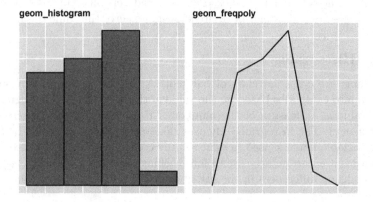

Histograms involving multiple categories benefit greatly from using *posi-tion = "dodge"*. Notice how in the graph below using *fill = group* results in a portion of the second group (in black) being trapped on top of the stack of the third group. This isn't helpful to the reader. However, this can be easily fixed by using *position = "dodge"*:

```
ggplot()+
   geom_histogram(data = demo2, aes(x = x, fill = group),
               position = "dodge")
```

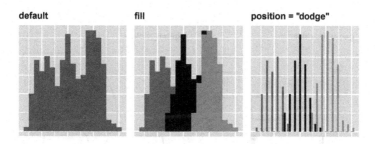

7.4.5   Box plots

A box plot is useful way of understanding differences between categories in terms of their quartiles. The centre line of an individual box plot indicates where the median is (which is the same as the 50th percentile = 2nd quar-tile). The lower and upper margins of the box represent the 25th and 75th percentiles (the 1st and 3rd quartiles). The line extending from the box are called whiskers. Whiskers represent the distance from the box to the largest value within the 1.5 x the inter-quartile range. Values greater than this dis-tance, if present, will appear as dots. Such dots represent outliers.

A violin plot is very similar to a box plot. A violin plot can be thought of as a symmetrical histogram lying on its side. Just like a box plot, it shows the spread of the data, but in addition, its shape indicates where the data points lie.

- The **geom_boxplot**() function takes the form:

```
ggplot()+
  geom_boxplot(data = demo1, aes(x = group1, y = y))
```

- The **geom_violin()** function takes the same form:

```
ggplot()+
  geom_violin(data = demo1, aes(x = group1, y = y))
```

Note: in both cases we use aesthetic mapping $x = group1$ to separate out groups on the x-axis.

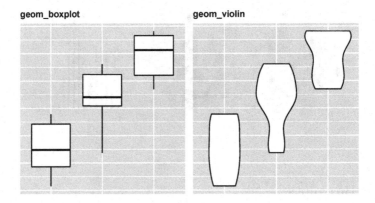

7.4.6   Polygons

Polygons can be imagined as a data frame in which the first and last data points have exactly the same coordinates. This is exactly the same as drawing a shape with a pencil. We can only finish the shape once we get back to the the same point where we originally started (Figure 7.2).

Most maps are based on polygons. These polygons are based on the joining of points. But rather than just having a few points (like the pentagon in our example) real maps are built upon connecting large number of points. By using the **geom_polygon()** function in combination with R packages specialising in reading spatial (map) data we can make maps in a ggplot (see Chapter 12 for more advanced approaches):

- In **geom_polygon()** the function takes the form:

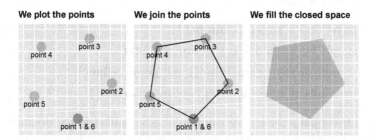

FIGURE 7.2
How a polygon is formed

```
ggplot()+
    geom_polygon(data = waihi, aes(x = long, y = lat,
                                   group = group))
```

Often map data will have its x and y coordinates renamed **long** and **lat**. In addition, map data converted into a data frame will usually have a column labelled **group**. The **group** variable contains information which makes sure that boundaries are connected in the correct order. If the *group = group* aesthetic is left out the polygon becomes a crazy abstract mess.

geom_poly()

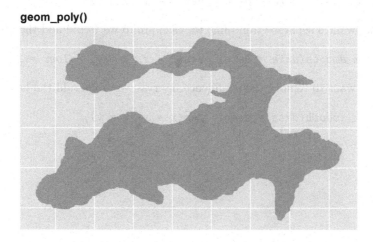

7.4.7   Other common geoms

As we progress in our analysis of our data we often need to add other geoms
to the ggplot, such as:

  − trend lines e.g. **geom_smooth()**.

  − simple linear models e.g. **geom_abline()**.

  − thresholds e.g. **geom_hline()** and **geom_vline()**.

  − measures of uncertainty e.g. **geom_errorbar()**.

All of these plots are just an additional geom added to the original ggplot
code. You'll need to remember to add a '+' to the original lines of code to have
them included. All of the following examples build upon the base of:

```
ggplot()+
  geom_point(data = demo1, aes(x = x, y = y))
```

As a starting point we might want to see if there any evidence of a trend in
the data. Often we will start this process with the **geom_smooth()** function.
This function plots a trend line (a smoother) based on a localised average.
It also adds a 95% confidence interval to the smoother. Such smoothers work
well on large data sets with hundreds or thousands of points. If you are using
a smoother with a small number of data points then you need to be very
cautious. Smoothers are not an answer by themselves.

• In **geom_smooth()** the function takes the form:

```
geom_smooth(data = demo1, aes(x = x, y = y))
```

If we want a separate trend line for each group we would add the layer:

```
geom_smooth(data = demo1, aes(x = x, y = y, colour = group1))
```

If we want to remove the confidence interval we would add the layer:

```
geom_smooth(data = demo1, aes(x = x, y = y),
            fill = NA)
```

geom_smooth

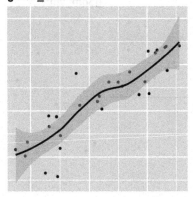

- **geom_abline()** is useful for plotting a simple linear model. In most cases you probably will manually add the intercept and slope (see Chapter 15). If no reference is being made to the data frame, the *data* attribute and the **aes()** function can be removed:

```
geom_abline(intercept = 0.6870, slope = 0.9202)
```

geom_abline

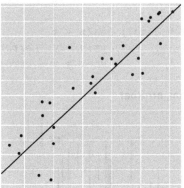

- **geom_hline()** draws a horizontal line across the plot panel from a *y* intercept:

```
geom_hline(yintercept = 9)
```

- In a similar manner **geom_vline()** draws a vertical line across the plot panel from a *x* intercept.

```
geom_vline(xintercept = 10)
```

geom_hline   **geom_vline**

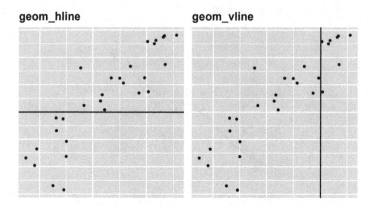

- **geom_errorbar()** can add error bars aligned with the x-axis or y-axis. These errors are not automatically calculated but must be created by the user and made into a variable. If the errors are symmetrical (e.g. **error** in the example below) then a single variable can be added to, or subtracted from our variable of interest to show the minimum and maximum values. If the errors are not symmetrical then two variables will be required: corresponding to the minimum and maximimum values of the error (e.g. *ymin* and *ymax*).

```
geom_errorbar(data = demo1, aes(x = x,
                          ymin = y - error,
                          ymax = y + error ))
```

geom_errorbar

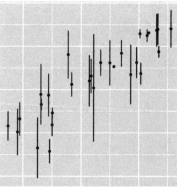

7.5   How to save a ggplot

There are a number of ways to save a ggplot:

1. The **ggsave()** function will save the last graph displayed. The function can save graphs as common file types (e.g. eps, jpeg, pdf, tiff, png). The only argument we have to supply to this function is the file's name including its file extension (e.g. png, jpg, pdf etc.):

```
ggsave("filename.png")
```

Unless we specify the size of the graph it will automatically make the image a default size (typically 7×7 inches). We can customisein RStudio this is determined by the size of the Plots pane). We can customise the size of the image using *width*, *height* and *units* arguments (unit options include "in", "cm", and "mm"):

```
ggsave("filename.png",  width = 10, height = 10,
       units = "cm")
```

2. In the default script editor we can also click on the R graphics window and go to `File > Save` as and then choose our file type.

3. Alternatively, using RStudio we can choose `Export > Save as Image` option from the plot pane. This produces a 'Save Plot as Image' window. This window has the advantage that it allows us to see the consequences of resizing the image (via the `Update Preview` button) before saving.

7.6   Recommended resources

- Hadley Wickham and Garret Grolemund (2016) R for Data Science. O'Reilly Media, Sebastopol. A free online version is available at: `https://r4ds.had.co.nz`.

7.7 Summary

- The `ggplot2` package within the `tidyverse` allows us to make most types of graphs from a basic code structure with only minor variations.

- In order to make a ggplot graph we need to know about:

    - geoms (the names of the graph types).
    - aesthetics (the variables/columns of data we want to graph).

8

Customising a ggplot

In this chapter we will learn about:

- The families of layers.
- How to make a ggplot more readable.
- The parts of a ggplot graph.
- How to customise the look of a ggplot.

8.1 Why customise a ggplot?

A good graph is usually the starting point for communication about data. While the default settings of ggplot are often fine – ultimately, it is a human who has to understand the graph. Bad combinations of colours, shapes, and line type can make a graph impossible to understand. Additionally, if our graph is going to be published by an organisation, whether it is for a poster, a webpage, social media, or a formal report chances are we are going to have to follow organisation guidelines to ensure the look of the graph is in keeping with their protocols or branding. For these reasons we need to be able to customise our ggplots.

8.2 The packages

In this chapter we will use the `tidyverse` and the `scales` packages. The `scales` package is used to make axes more readable. As we haven't used `scales` before it will have to be installed (it is available on CRAN). Let's load these two packages:

```
library("scales")
library("tidyverse")
```

8.3   The data

For this chapter we will use the `demo1` from the `condev` package. Start by loading `condev` and `demo1` data set:

```
library("condev")
data(demo1)
```

8.4   Families of layers

When we are making a ggplot it is helpful to think of 7 families of layers. We always need two layers to make a ggplot:

1. **ggplot()**: sets up the coordinate system.

2. **geom_***()**: determines the type of graph.

But in addition to these there are another 5 families of layers that can help us alter our ggplot:

1. **scale_***()**: changes the axes and legends.

2. **coord_***()**: changes the coordinate system.

3. **facet_***()**: splits a ggplot into multiple graphs.

4. **labs()**:   changes the labels.

5. **theme()**: changes the look of the graphic features.

While it is helpful to understand this system I have organised this chapter by the type of issue you might face. I only cover the most common types of issues. I recommend searching `stackoverflow.com` or the books listed at the end of this chapter for further help.

8.5   Aesthetics properties

The simplest way to improve a plot is by choosing the aesthetics in the **geom_***()** function which make the graph look better. Figure 8.1 gives the main options associated with ggplot aesthetics. A description of these aesthetic elements is given below:

- *size* controls the size of the aesthetic element (e.g. point or line).

- *colour* controls the interior and exterior colour of shape codes ≤ 20. As well as the borders of bars, boxes, and shape codes ≥ 20.

- *fill* controls the interior colour of non-points (e.g. bars and boxes) and shape codes ≥ 20.

- *alpha* controls the transparency via a decimal between 1 (no transparency) to 0 (full transparency).

- *fontface* controls choice of font type.

- *linetype* controls choice of line type.

Occasionally, we will want to distinguish a group without necessarily changing its look (e.g colour, fill etc). In such cases we can use the aesthetic *group*.

Note: The ggplots presented in this chapter have been slightly stylised for publication. If you use the argument *size=5* within the **geom_point()** function your ggplots will more closely resemble the examples given.

8.5.1 Settings aesthetics

There are broadly two ways to change the look of an aesthetic:

1. Give the aesthetic a value – outside of the **aes()** – this will apply to all data points e.g. for colour:

```
ggplot()+
  geom_point(data = demo1, aes (x = x, y = y),
             colour = "red")
```

or shape:

```
ggplot()+
  geom_point(data = demo1, aes (x = x, y = y),
             shape = 2)
```

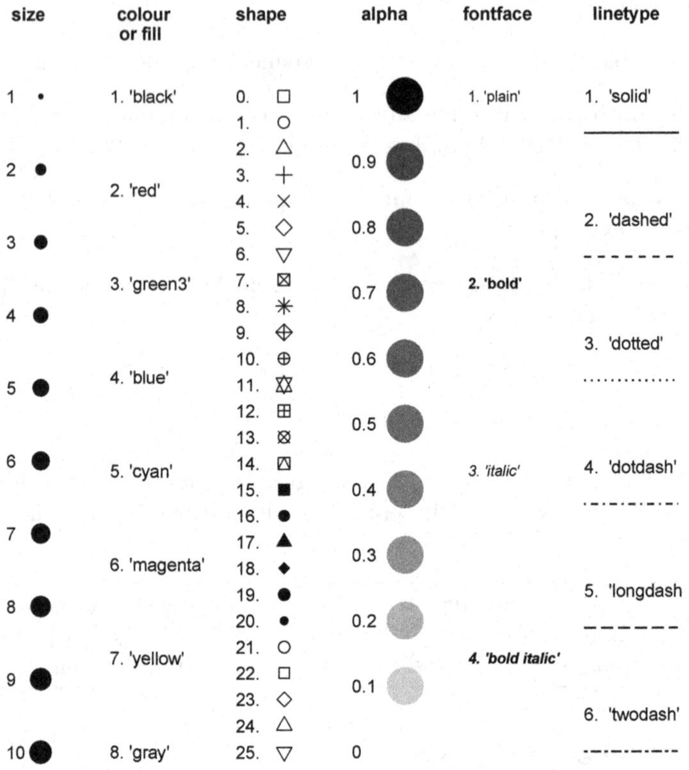

size	colour or fill	shape	alpha	fontface	linetype
1 ·	1. 'black'	0. ☐	1	1. 'plain'	1. 'solid'
		1. ○			
2 ●		2. △	0.9		
	2. 'red'	3. +			
		4. ×			
3 ●		5. ◇	0.8		2. 'dashed'
		6. ▽			
	3. 'green3'	7. ⊠	0.7	2. **'bold'**	
4 ●		8. ✳			
		9. ⊕			3. 'dotted'
		10. ⊕	0.6		
5 ●	4. 'blue'	11. ⋈			
		12. ⊞			
		13. ⊠	0.5		
6 ●	5. 'cyan'	14. ◩		3. *'italic'*	4. 'dotdash'
		15. ■	0.4		
		16. ●			
7 ●		17. ▲	0.3		
	6. 'magenta'	18. ◆			
		19. ●			5. 'longdash
8 ●		20. •	0.2		
	7. 'yellow'	21. ○		4. ***'bold italic'***	
9 ●		22. ☐	0.1		
		23. ◇			6. 'twodash'
		24. △			
10 ●	8. 'gray'	25. ▽	0		

FIGURE 8.1
Aesthetic options for a ggplot

2. Manually change the look using **scale_***_manual()** where the middle name of the function is the type of aesthetic you are changing e.g. for the colour of groups:

```
ggplot()+
   geom_point(data = demo1, aes (x = x, y = y,
                                 colour = group1))+
   scale_colour_manual(values = c("red", "yellow",
                                  "black"))
```

or for the shape of groups:

```
ggplot()+
  geom_point(data = demo1, aes (x = x, y = y,
                                 shape = group1))+
  scale_shape_manual(values = c(3, 4, 5))
```

If the aesthetic colour is related to a continuous variable, then ggplot will vary the intensity of the colour along a blue spectrum by default. One way to change this spectrum is to use the **scale_colour_gradient**() function to set different colours at the extremes (low and high ends):

```
ggplot()+
  geom_point(data = demo1, aes (x = x, y = y,
                                 colour = y))+
  scale_colour_gradient(low = "yellow", high = "black")
```

8.5.2 A quick note about colour

There are eight default colours available in ggplot. However, there are 657 named colours available in R. The names of these colours are available by:

```
colors()
```

Another easy way to assign a particular colour is to use the colour's hex code. A hex code is a way of specifying colour using a 6 element code made up of numbers and letters. There are lots tools on the internet for finding the names of these codes. Simply search "hex colour codes". Consequently, the following would all result in the colour red: "red", "#FF0000", or 2 (the ggplot code for red – see Figure 8.1).

8.5.3 Using aesthetics to distinguish groups

Aesthetics can be used to help distinguish different data groupings (e.g. categories of a variable). While colour is good to distinguish groups of points on a screen it is worth while remembering that around 8% of men and 1% of women are colour blind. Also if the plot is going to be printed in black and white the points will be recognised by their shade, not by their colour. While shape can be helpful it can be also visually overwhelming. Using combination of shape and colour at the same time is often a better option (Figure 8.2):

```
ggplot()+
  geom_point(data = demo1, aes(x = x, y = y, colour = group1,
                              shape = group1))
```

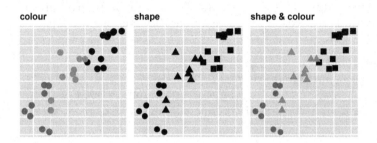

FIGURE 8.2
Using combinations of colour and shape to distinguish groups

8.5.4   Using faceting to distinguish groups

An alternative to just using colour and shape is to use faceting. Faceting gives each category of a variable its own plot. Because facetting can take two variables at once it is helpful for finding patterns that relate to different combinations of categories. It also can help remove issues associated with colour blindness or black and white printing. The **facet_grid()** function uses a formula which places the plots by rows and columns:

```
facet_grid( . ~ group1)      # group1 are the columns
facet_grid(group1 ~ . )      # group1 are the rows
facet_grid(group1 ~ group2)  # group1 are the rows,
                             # group2 are the columns
```

In this way we can use facetting to produce a row of graphs (Figure 8.3) or a grid of graphs (Figure 8.4).

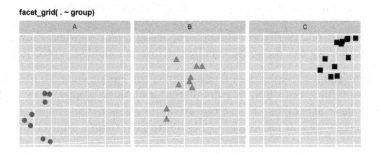

FIGURE 8.3
Facetting with columns only

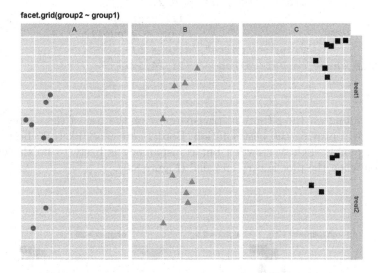

FIGURE 8.4
Facetting with both rows and columns

8.6 Improving crowded graphs

If we are working with larger data sets sometimes our plots become very crowded and data points start to get hidden. In such situations, it becomes unclear whether a point at one location is just single data point or multiple points of the same value. There are two ways to overcome this:

- Make data points slightly transparent using *alpha* so when they overlay each other we can recognise this by their darker shade (see section 8.7).

- Use the shapes greater than number 20 on the ggplot scale. Unlike normal shapes (which only have a colour attribute) these shapes have both *fill* and *colour* attributes. The colour attribute refers to the shape's border – this means we can use *colour = "white"* to give them a white border.

Below is an example of using shapes greater than number 20 with a white border (and a larger size) to make a clearer graph:

```
ggplot()+
  geom_point(data = demo1, aes(x = x, y = y,
                               fill = group1,
                               shape = group1),
             colour = "white", size = 5)+
  scale_shape_manual(values = c(21,24,22))
```

Depending on the situation, transparency or the use of the shapes with borders will provide the best solution. Borders are most likely to be helpful when a plot is slightly crowded but transparency is likely to be the better solution when the plot is very crowded (Figure 8.5).

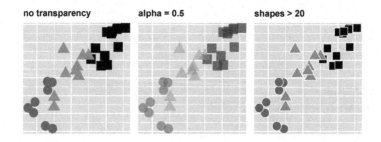

FIGURE 8.5
Improving crowded graphs with transparency and shapes values 20 and greater

8.7   Overlaying

One of ggplot's great features is the ability to overlay various different geoms to create more complex visualisations. However, to make this work, the aesthetics of the different geoms should be of the same data type[1]:

```
ggplot()+
   geom_boxplot(data = demo1, aes(x = group1, y = y))+
   geom_point(data = demo1, aes(x = group1, y = y))
```

The danger with this approach is that one plot can obscure another. There are two common strategies:

1. Make the layers transparent using the *alpha* argument where *alpha* is a decimal between 1 (no transparency) to 0 fully transparent (e.g. *alpha* = 0.5).

2. Adding a jitter to points. A jitter is a random offset (noise) to the point which can be controlled by *width* and *height* arguments[2]. Adding a small amount of noise usually helps separates points which are obscuring each other. This can be done in two ways:

   By using *position* = **position_jitter()** in **geom_point()**:

```
geom_point(data = demo1, aes(x = group1, y = y),
              position = position_jitter(width = 0.1,
                                         height = 0))
```

   or by using the **geom_jitter()** function:

```
geom_jitter(data = demo1, aes(x = group1, y = y),
              width = 0.1,  height = 0)
```

This technique is particularly helpful when overlaying geoms which summarise a group's distribution e.g. **geom_boxplot()** and those which show the individual observations e.g **geom_point()** (see Figure 8.6).

---

[1]in the example that follows we can force an error by changing the *y* variable of one geom to a factor using y = as.factor(y)

[2]if you want to ensure there is no jitter on an axis you must set one of these arguments to zero

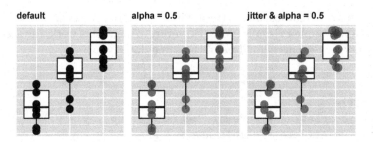

FIGURE 8.6
Overlaying two ggplot layers with transparency and jittering

8.8 Labels

Figure 8.7 outlines the main labels associated with a ggplot graph. At a minimum, axis labels and some form of explanation of the graph (either in the form of a title or caption) are usually expected.

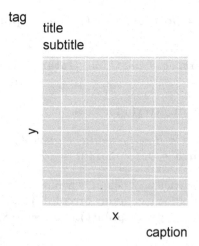

FIGURE 8.7
The main labels of a ggplot

More rarely, a tag (such as a number or letter) can also be added – but this is usually only used when you are producing multiple graphs of a similar

nature on a single page. Labels can be given to all these elements using the
labs() function. Only the elements you name will be given a label e.g.:

```
ggplot()+
  geom_point(data = demo1, aes(x = x, y = y))+
  labs(title = "My scatter graph",
       subtitle = "Default theme",
       x = "explanatory variable",
       y = "response variable")
```

Which is the same as:

```
ggplot()+
  geom_point(data = demo1, aes(x = x, y = y))+
  labs(title = "My scatter graph")+
  labs(subtitle = "Default theme")+
  labs(x = "explanatory variable")+
  labs(y = "response variable")
```

And results in:

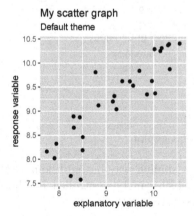

Whether you choose to define the labels using a single function, or as a set
of separate functions is entirely your choice. Do which ever you find easier to
read. There are also a variety of shorthand functions for certain label features:

```
ggtitle("My scatter graph")      # title
xlab("explanatory variable")     # x label
ylab("response variable")        # y label
```

8.9   Using the theme() function

A ggplot graph is made up of a number of different graphical features. All of these can be controlled by the use of the **theme()** function[3]. However, generally there are only a few key features that need to be controlled (Figure 8.8).

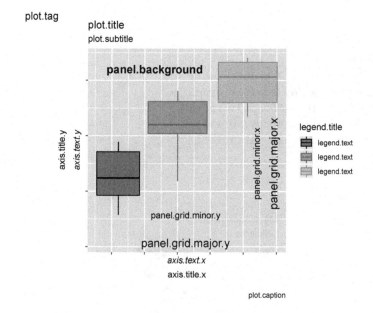

FIGURE 8.8
The main theme elements of a ggplot

Facets are controlled in a similiar way but additionally have strips on the top and side of the plots which can be controlled by the `strip.text` and `strip.background` features of **theme()** (Figure 8.9).

---

[3]Full list at `https://ggplot2.tidyverse.org/reference/theme.html`

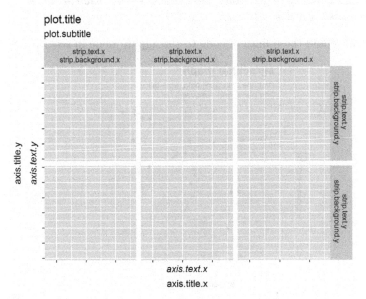

FIGURE 8.9

The additional theme elements of a facetted ggplot

8.9.1 The 4 elements

Each feature in **theme()** can be controlled by 1 of 4 functions which control the graphical elements:

1. **element_blank()**: removes the feature entirely.

2. **element_line()**: controls line features.

3. **element_rect()**: controls rectangular features.

4. **element_text()**: controls text features.

We can control each element by changing the attributes within it. If we want to turn off an element we use **element_blank()**. The most common functions used are **element_text()** and **element_blank()**.

When a theme feature is separated by more than 2 periods e.g. *axis.title.y* or *panel.grid.major.x* this represents a sublevel of the feature. If we wanted to change all sublevels of a feature we can just use the core name of the feature (e.g *panel.grid*). For example, if we want to make the all panel grid lines disappear and make the plot title text bigger, italic, and bolder we would add the lines (notice that the feature is called face in theme() not fontface):

```
theme(plot.title = element_text(size = 14, face = 4),
      panel.grid = element_blank())
```

Which is the same as:

```
theme(plot.title = element_text(size = 14, face = 4))+
theme(panel.grid = element_blank())
```

When we add these lines of code to the ggplot made in section 8.8 we get:

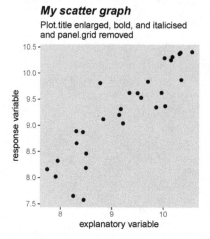

8.9.2 Rotating axis text

The default position for axis text in a ggplot is horizontal. Often we will want to rotate an axis label – this is especially true if the axis text is particularly long. This can be accomplished using the *angle* argument:

```
theme(axis.text.x = element_text(angle = 90))
```

When add these lines of code to the ggplot made in section 8.8 we get:

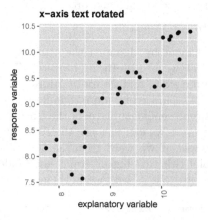

We now see that the x-axis text doesn't line up with the tick marks. This is because the text is not justified correctly. To fix these issues we use the *hjust* argument (for horizontal justification) or *vjust* argument (for vertical justification):

- − 0 = left justified
- − 0.5 = centre justified
- − 1 = right justified

In our example, because we have just rotated the x-axis we will need to use *vjust* to recentre the text:

```
theme(axis.text.x = element_text(angle = 90, vjust = 0.5))
```

When we add these lines of code to the ggplot made in section 8.8 we get:

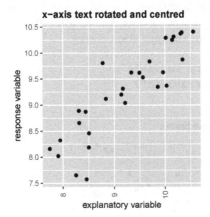

8.9.3 Spacing between axis and graph

Sometimes we will want to alter the spacing between the axis title and text labels. We can control this through the **margin()** function. This sets the values of the margin on **T**op, **R**ight, **B**ottom, and **L**eft sides. You can remember this pattern as "TRouBLe". For example, if we want to shift the graph to the right away from the y-axis title we would use:

```
theme(axis.title.y =
        element_text(margin =
                    margin(t=0, r=30, b=0, l=0, "mm")))
```

When add these lines of code to the ggplot made in section 8.8 we get:

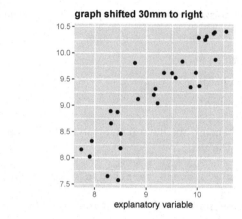

Alternatively, if we want a little more empty space around an axis title we can insert a line break before the x-axis title or after the y-axis title. We do this by inserting '**\n**' within the quotation marks:

```
labs(x = "\nexplanatory variable",
     y = "response variable\n")
```

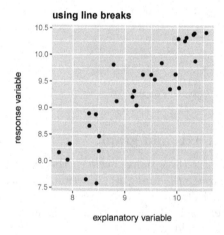

8.9.4   In-built themes

There are also 8 in-built themes which can be used (Figure 8.10). By using these you can quickly produce a good looking graph for a report with a short line of code referencing the theme function e.g. **theme_minimal()** function:

```
ggplot()+
  geom_point(data = demo1, aes(x = group1, y = y))+
  theme_minimal()
```

Note: if you add an in-built theme after you have manually altered a ggplot with the **theme()** function the in-built theme will override your **theme()** settings. The way around this is make sure your manual theme settings occur after the code for the in-built theme.

8.10 Controlling axes

The main ways axes can be controlled in ggplot is through the **scale_***()** and **coord_***()** family of functions. I also recommending installing the `scales` package to improve the look of axes when dealing with large numbers.

8.10.1 Tick marks

In ggplot the tick marks and the text associated with them are controlled through the **scale_*axis*_*type*()** function where:

- where ***axis*** is either **x** or **y**.
- where ***type*** is either **discrete** (for categorical variables) or **continuous** (for continuous variables).

The most common arguments used with these functions are:

- *breaks* = the separation between tick marks on that axis.
- *labels* = the labels used for each tick mark on that axis.

We can alter the breaks by giving a sequence of divisions to ggplot using the **seq()** or **c()** functions:

```
scale_y_continuous(breaks = seq(from = 7.5, to = 10.5, by = 1))
```

We can even assign text labels to a continuous axis so long as the length of the breaks and labels match (Figure 8.11):

```
scale_x_continuous(breaks = c(8,9,10), labels = c("a","b","c"))
```

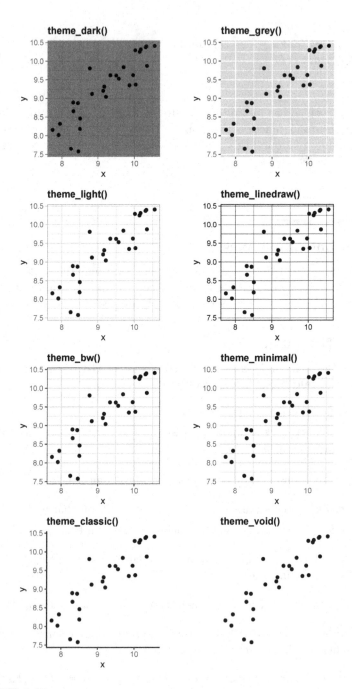

FIGURE 8.10
In-built ggplot themes

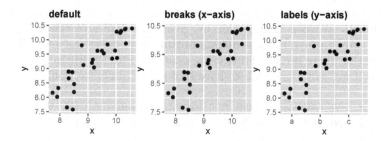

FIGURE 8.11
Setting breaks and labels. Notice the use of letters as labels (right most ggplot)

8.10.2 Axis limits

The axis limits of a ggplot are controlled separately for each axis e.g.:
scale_x_continuous() and **scale_y_continuous()** function:

```
scale_x_continuous(limits = c(8, 9))
scale_y_continuous(limits = c(8, 10))
```

The same result can be achieved using the short hand method using the **xlim()** and **ylim()** functions (see Figure 8.12):

```
xlim(8, 9)
ylim(8, 10)
```

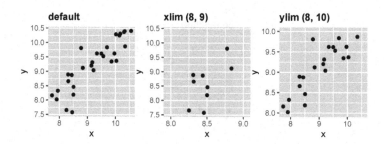

FIGURE 8.12
Controlling axis limits with xlim() and ylim()

8.10.3 Forcing a common origin

The default ggplot settings mean the the x-axis and y-axis don't share a common origin because there is a slight separation between axes (known as padding). We can change this by giving the limits of the axes to the **scale_x_continuous()** and **scale_y_continuous()** functions and telling the function that there is no padding using *expand = c(0, 0)* (Figure 8.13). So for example if we want our graph to have zero as the origin we would add:

```
scale_x_continuous(limits = c(0, 11), expand = c(0, 0)) +
scale_y_continuous(limits = c(0, 11), expand = c(0, 0))
```

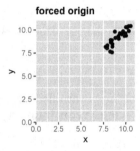

FIGURE 8.13
Forcing a common origin with scale_x_continuous() and scale_y_continuous()

8.10.4 Flipping axes

Axes can be easily flipped by adding the **coord_flip()** function as a separate empty layer (Figure 8.14):

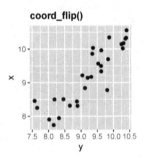

FIGURE 8.14
Flipping axes with coord_flip()

8.10.5 Forcing a plot to be square

The default plot of ggplot is normally as wide as your plot window. To force
a plot to be square we add the **coord_equal()** function as a separate empty
layer (Figure 8.15):

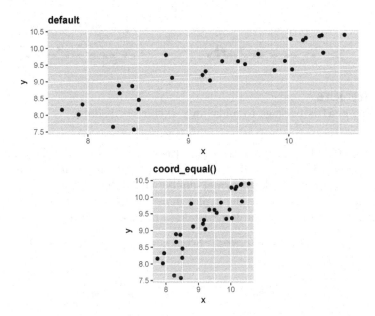

FIGURE 8.15
Plot shape: default vs coord_equal()

8.10.6 Log scales and large numbers

When the values in our data vary by orders of magnitudes it is often hard
to visualise the data points using the default ggplot settings. If the axis values are particularly large they will be expressed in exponential notation. In
exponential notation numbers are expressed as a number between 1 and 10
followed by an appropriate power of 10. For example:

$3{,}000{,}000 = 3 \times 10^6$ is written as 3e + 06

Such notation can be confusing and hard to read. Using the `scales` package
we can force the notation back into the normal numeric format and have it
expressed using commas using the following code (Figure 8.16):

```
scale_y_continuous(label = comma)
```

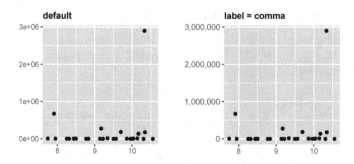

FIGURE 8.16
Using the `scales` package to make an y-axis more readable

Further to this, we can log transform the y-axis using the **scale_y_log10()** function to distinguish the data more clearly. Not only can we change the exponential notation back to normal numbers by including the *label = comma* argument, but we can make the ggplot more readable by setting the number of y-axis breaks by using the **log_breaks()** function (Figure 8.17):

```
scale_y_log10(label = comma, breaks = log_breaks(n = 7))
```

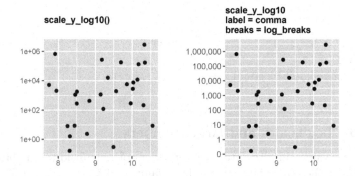

FIGURE 8.17
Using the `scales` package to make log transformed ggplots more readable

8.11   Controlling legends

There are 4 standard positions for a ggplot legend: *top, right, bottom, and left.* To control the position of a legend we use the *legend.position* argument of the **theme()** function:

```
theme(legend.position = "top")
```

All legends can be turned off using:

```
theme(legend.position = "none")
```

Legends in ggplot are based on the aesthetics used. If no special aesthetic elements (e.g. *colour*, *fill* etc) are used then no legend will be produced. Individual legends can be turned off by using the **guides()** function with the argument **** = FALSE* where ***** is the name of the aesthetic we want to turn off e.g. if we wanted to hide just the legend associated with colour of the groups:

```
guides(colour = FALSE)
```

8.12   Recommended resources

- Hadley Wickham and Garret Grolemund (2016) R for Data Science. O'Reilly Media, Sebastopol. A free online version is available at: https://r4ds.had.co.nz.

- Claus O. Wilke (2019) Fundamentals of Data Visualization. O'Reilly Media, Sebastopol. A free online version is available at: https://serialmentor.com/dataviz/.

8.13   Summary

When we are making a ggplot it is helpful to think of 7 families of layers:

1. **ggplot()**: sets up the coordinate system.

2. **geom_***()**: determines the type of graph.

3. **scale_***()**: changes the axes and legends.

4. **coord_***()**: changes the coordinate system.

5. **facet_***()**: splits a ggplot into multiple graphs.

6. **labs()**: changes the labels.

7. **theme()**: changes the look of the graphic features.

9

Data wrangling

In this chapter we will look at data wrangling – getting your data into the form you want. We will cover:

- Subsetting.

- Transforming.

- Tidying.

- Summarising.

- Ordering.

- Joining.

9.1 What is data wrangling?

Typically, in conservation and development projects the person analysing the data is not the same person who collected the data, entered the data, or planned the study. Consequently, the person analysing the data will often have to rearrange it to meet their needs. Often this means subsetting, modifying, summarising, and ordering the original data – known as data wrangling. For example, we might want to get information for fifty villages from a government data set which contains information on thousands of villages. In addition, graphing and modelling usually require the data in particular layout, as a result we will need to have the skills to change the format of our data to match. Manually reformatting small data sets maybe an option but there is the risk of making making errors, however, with large data sets it may be impractical – in which case we will need R's help. This chapter will demonstrate some of the most useful functions and show how we can wrangle data with a few lines of code.

9.2   The packages

While base R can be used to wrangle data, the `tidyverse` set of packages
makes data wrangling easier. Consequently, we will focus on using the functions within the `tidyverse`. Another useful package that we will also use is
`janitor` which is designed for cleaning and summarising data. These two packages work well together. These packages will be used throughout this chapter
so load them now:

```
library("janitor")
library("tidyverse")
```

As the `tidyverse` package is actually a collection of related packages in
this chapter I indicate the package associated with each of the key functions.
This should make it easier for the reader to find additional help if it is required.

9.3   The data

For this chapter we will use two example data sets (**attendees** and **hunt**)
from the `condev` package. Start by loading the `condev` package and the data
sets below:

```
library("condev")
data(attendees)
data(hunt)
```

9.4   Pipes - `tidyverse(magrittr)`

When we wrangle data we often have to apply a sequence of functions to the
data set to get it into the form we need. Let's consider a very simple example
in which we want to get the mean value of the object `test` rounded to 1
decimal place:

```
test <- c(2,2,3)
```

There are a couple of ways to approach this:

1. We could nest the functions:

```
result <- round(mean(test),1)
result

[1] 2.3
```

The downside to this approach is that our code rapidly becomes unreadable if we need to nest more than a couple of simple functions.

2. Apply the functions one at a time and each time overwrite the object using intermediate steps:

```
result <- mean(test)
result <- round(result,1)
result

[1] 2.3
```

While this approach is definitely easier to read than the nested approach there is a lot of unnecessary repetition of the object's name.

3. Use a pipe (%>%) from the **tidyverse** package. Pipes passes the output of the function on to the next function as its main argument. In the following example, we pipe the object **test** into the **mean()** function and then its output into the **round()** function:

```
result <- test %>%
  mean() %>%
  round(1)
result

[1] 2.3
```

The advantage with using pipes is:

- We don't have to write the name of the object within the functions.
- We don't have to keep overwriting our object.
- We can avoid nesting functions.

Not all packages and functions will support pipes. However, pipes will work with any function or package contained within the `tidyverse`. In this chapter we will not save the output of a pipe as an object (unless we have to reuse the object in a later example). However, in the normal course of writing code you probably will want to save them as objects so that they are stored in R's memory. This way you can refer back to them as you need. Also, if it helps you avoid mistakes, it is totally okay to breakup the pipes into shorter blocks of code.

9.5   Tibbles versus data frame

So far in this book we have been using data frames. However, prior to R version 4.0 (released in April 2020) it had been a little difficult to change text held in data frames. This was because in earlier versions columns with text (i.e. characters) were automatically changed into factors when they were imported into R[1].

The advantage of using factors is that they can show us when certain categories are absent in the data, and they help speed up calculations, but the downside is they are harder to change because extra information is coded into pre-defined levels. As a result, a simplified form of data frame was created, called a tibble, to make changing text easier in the `tidyverse`. Tibbles never change the original format of the data (regardless of R version). As some `tidyverse` functions will always result in a tibble it is good to have some understanding off them.

We can make a tibble in the same way as a data frame but use the **tibble()** function rather than the **data.frame()** function. For example, if we wanted to change the existing `hunt` data frame to a tibble we would use the **as_tibble()** function:

```
our.tibble <- as_tibble(hunt)
our.tibble
```

```
# A tibble: 20 x 6
   hunter.ref    age village    distance hours catch.weight
```

----

[1]in R version 4.0 *stringsAsFactors* = *FALSE* became the default for importing data – see Chapter 13 for more information

<fct>	<int>	<fct>	<dbl>	<dbl>	<dbl>
1 H1	49	Anota	1.73	6.35	6.71
2 H2	29	Takendu	2.13	6.29	8.29
3 H3	35	Nugi	2.99	5.09	7.19
4 H4	22	Anota	4.02	4.21	8.62
5 H5	60	Nugi	2.54	5.36	4.92
6 H6	18	Takendu	2.37	5.47	8.72
7 H7	43	Nugi	2.67	5.4	6.81
8 H8	52	Maniwavie	2.27	5.64	5.57
9 H9	56	Takendu	2.6	5.15	5.95
10 H10	20	Anota	2.71	5.54	8.99
# ... with 10 more rows					

Usually when we run `our.tibble` the output shows just the first 10 rows of the tibble as well as the data type directly under the column name. However, if there less than 20, rows all will be shown. The data types stored in a tibble are named slightly differently to that of data frames, and are given below:

1. `<dbl>`: double[2] – numbers with decimals e.g. 4.7, 15.57. This is equivalent to 'numeric' in data frames.

2. `<int>`: integer – whole numbers (including zero).

3. `<chr>`: character – a string of numbers and/or letters enclosed in quotation marks.

4. `<fct>`: factor – a categorical variable with a set number of levels.

5. `<date>`: date – a date format recognised by R.

6. `<lgl>`: logical – one of two values: TRUE or FALSE.

Tibbles also only show a portion of the data. This is done to prevent the console window from becoming unreadable. If we want to see more of the data we can use the **print()** function. If we want more rows we use the n argument:

```
print(our.tibble, n = 17) # first 17 rows
print(our.tibble, n = Inf) # all rows (Inf = infinity rows)
```

If we had a very wide tibble and wanted more to see more columns we could use the *width* argument. However, it is probably easier to view such a tibble with the **View()** function in RStudio or the **fix()** function in base R.

---

[2]short for double-precision floating-point format

9.6   Subsetting

Often we are only interested in using part of a data set. For example, we
might have access to a large government data set but we only want a subset of
information that relates to our project site e.g. a few columns of the data or
data that relates to specific observations or categories. As a result, we would
want to subset the data. While we could use base R to subset the data (see
Chapter 5) the `tidyverse` has a set of functions which makes this easy.

9.6.1   select() - `tidyverse(dplyr)`

The **select()** function allows us to work with a subset by columns. The **se-
lect()** function doesn't change the format of the data set:

```
attendees %>%
  select(village, name)
```

The top six lines of the output are shown below:

```
    village           name
1 Lamaris    Nathan Rivera
2 Takendu      Bryant Han
3    Nugi  Yangzi Jameson
4   Anota   Zhening Chang
5 Takendu      Tuong Pham
6   Anota Lewis Donthinani
```

If we wanted, we could remove the same columns by adding a negative sign
in front of the selection.

```
attendees %>%
  select(-c(village, name))
```

The top six lines of the output would then become:

```
  gender  course training.dates
1   male vanilla       8/10/2018
2   male vanilla       8/10/2018
3   male vanilla      12/10/2018
4   male vanilla       8/10/2018
5   male vanilla       8/10/2018
6   male vanilla       8/10/2018
```

9.6.2 filter() - tidyverse(dplyr)

The **filter()** function allows us to subset a data set by its values. In order to use this function we need to be able to use logical operators (the most common ones are given in Table 9.1). For example, we might choose to filter a data set to check information associated with specific values or categories.

TABLE 9.1
Common logical operators in R

Code	Description
<	less than
<=	less than or equal to
>	greater than
>=	greater than or equal to
==	exactly equal to
!=	not equal to
|	or
&	and

For example, we could filter the **attendees** data set to find our what training courses Tuong Pham attended:

```
attendees %>%
  filter(name == "Tuong Pham")

  gender village      name    course training.dates
1   male Takendu Tuong Pham   vanilla      8/10/2018
2   male Takendu Tuong Pham  forestry      5/10/2018
3   male Takendu Tuong Pham watersheds     6/10/2018
```

We can also make multiple requests. Note that the variable (column) name has to be repeated:

```
attendees %>%
  filter(name =="Tuong Pham" |
         name == "Jenaer Thapa")

  gender village         name      course training.dates
1   male Takendu   Tuong Pham     vanilla      8/10/2018
2   male    Nugi Jenaer Thapa     vanilla      8/10/2018
3   male Takendu   Tuong Pham    forestry      5/10/2018
4   male    Nugi Jenaer Thapa    forestry      5/10/2018
5   male Takendu   Tuong Pham watersheds      6/10/2018
```

The same approach can be taken with numerical values. In the example below we used the '&' sign:

```
hunt %>% filter(age >= 20 &
                age <= 25)

  hunter.ref age village distance hours catch.weight
1         H4  22   Anota    4.02  4.21         8.62
2        H10  20   Anota    2.71  5.54         8.99
```

We can also get the same result if we used a comma rather than the '&' sign.

9.7 Transforming

Often we will want to transform our data. Sometimes we will be wanting to create new columns derived from our existing ones, or create summary tables (e.g. for donor reports).

9.7.1 group_by() - `tidyverse(dplyr)`

The most important function in the `tidyverse` is the **group_by()** function. The function takes a data set and converts it into a set of groups, thereafter any functions are performed group by group. In the example below the **group_by()** function first separates the **attendees** data set by the 3 categories present in **course**, and then the 2 categories present **gender**. In this way the data is separated into 6 groups. The function does not change how the data looks, rather it changes how data acts, as we shall see later.

```
attendees %>%
  group_by(course, gender)
```

It is important to know that the data will continue to behave as a series of groups until the **ungroup()** function is used:

```
attendees %>%
  group_by(course, gender) %>%
  ungroup()
```

9.7.2 summarise() - `tidyverse(dplyr)`

As its name suggests the **summarise()** function summarises grouped data by one or more functions. As a consequence, we need to use it after the **group_by()** function. In the following example, we are summarising using the **n()** function[3]. The **n()** function counts the numbers of observations in each of the groups. This combination of functions is particularly valuable for creating summary tables from data sets where there are no numeric variables (e.g. attendance data). As expected from the number of categories in our groups (**course** = 3, **gender** = 2) we get a tibble with 6 rows as an output:

---

[3]this function only works within the **summarise()**, **mutate()**, and **filter()** functions

```
attendees %>%
  group_by(course, gender) %>%
  summarise(count = n())

# A tibble: 6 x 3
# Groups:   course [3]
  course     gender count
  <chr>      <chr>  <int>
1 forestry   female    36
2 forestry   male      56
3 vanilla    female    52
4 vanilla    male      90
5 watersheds female    43
6 watersheds male      57
```

We can also use the **summarise()** function to produce multiple new columns at the same time:

```
hunt %>%
  group_by(village) %>%
  summarise(average = mean(catch.weight),
            minimum = min(catch.weight),
            maximum = max(catch.weight),
            sample.size = n())

# A tibble: 5 x 5
  village   average minimum maximum sample.size
  <fct>       <dbl>   <dbl>   <dbl>       <int>
1 Anota        7.50    6.43    8.99           5
2 Lamaris      6.56    6.56    6.56           1
3 Maniwavie    5.77    5.18    6.69           5
4 Nugi         6.78    4.92    8.19           4
5 Takendu      7.09    5.95    8.72           5
```

9.7.3 mutate() - tidyverse(dplyr)

The **mutate()** function creates new columns. For example if we wanted to create a new column called `kg.per.hour` from the `catch.weight` and `hours` columns of `hunt` we would use:

```
hunt %>%
  mutate(kg.per.hour = catch.weight / hours)
```

The top three rows are:

	hunter.ref	age	village	distance	hours	catch.weight	kg.per.hour
1	H1	49	Anota	1.73	6.35	6.71	1.056693
2	H2	29	Takendu	2.13	6.29	8.29	1.317965
3	H3	35	Nugi	2.99	5.09	7.19	1.412574

The **mutate()** function is particularly useful after summarising grouped data to produce proportions and percentages:

```
attendees %>%
  group_by(course, gender) %>%
  summarise(count = n()) %>%
  mutate(proportion = count / sum(count))
```

```
# A tibble: 6 x 4
# Groups:   course [3]
  course     gender count proportion
  <chr>      <chr>  <int>      <dbl>
1 forestry   female    36      0.391
2 forestry   male      56      0.609
3 vanilla    female    52      0.366
4 vanilla    male      90      0.634
5 watersheds female    43      0.43
6 watersheds male      57      0.570
```

9.7.4 adorn_totals() - janitor

Most donors expect to see summary totals in report tables – but this can be a little time consuming in base R. Fortunately, we can make such totals with the janitor package with the **adorn_totals()** function. The position of the total is determined by using the arguments *"row"* or *"col"*. If totals are needed for rows and columns at the same time the argument *c("row", "col")* should be used. In order to make a total the **adorn_totals()** function requires at least one column in the data set to be numeric (other than the first column, which it assumes is always a reference variable). While the function can be used with raw data, it is probably most useful for presenting summary data:

```
attendees %>%
  group_by(course) %>%
  summarise(count = n()) %>%
  adorn_totals("row")

      course count
    forestry    92
     vanilla   142
  watersheds   100
       Total   334
```

9.8 Tidying

There are two basic formats of data: wide or long (Figure 9.1). Both formats can contain exactly the same data. The difference between the two is that wide formats have variables in different columns while long formats can have their variables and values in a restricted number of columns. Wide formats are often easier for people to read, but modelling and graphing in R often requires data to be in a long format. For example, a wide format data set with columns dedicated to males and females will need to be transformed to a single column with the genders being categories if we wanted to graph it as a ggplot. To transform between the two formats we need to use the functions **pivot_wider()** and **pivot_longer()**.

9.8.1   pivot_wider() - tidyverse(tidyr)

The **pivot_wider()** function requires name of the variable column (e.g. *names_from = gender*) and the name of the value column (e.g *values_from = count*) as arguments to change the data from a long format to a wide format:

```
wide <- attendees %>%
    group_by(course, village, gender) %>%
    summarise(count = n()) %>%
```

long

Person	Variable	Value
A	Variable1	16
B	Variable1	27
C	Variable1	18
A	Variable2	2.2
B	Variable2	2.1
C	Variable2	2.3
A	Variable3	178
B	Variable3	165
C	Variable3	143

Person	Variable1	Variable2	Variable3
A	16	2.2	178
B	27	2.1	165
C	18	2.3	143

wide

FIGURE 9.1
The difference between wide and long data formats

```
    pivot_wider(names_from = gender, values_from = count)
wide
```

As we can see the object we made is now in a wide format:

```
# A tibble: 15 x 4
# Groups:   course, village [15]
  course   village    female  male
  <chr>    <chr>       <int> <int>
1 forestry Anota           9    17
2 forestry Lamaris         3     4
3 forestry Maniwavie       4     8
4 forestry Nugi           12    12
5 forestry Takendu         8    15
6 vanilla  Anota          14    27
# ... with 9 more rows
```

9.8.2 pivot_longer() - `tidyverse(tidyr)`

The **pivot_longer()** function changes a wide format to a long format. This function is slightly more complicated than the **pivot_wider()** function. Here we need to give the function 3 pieces of information:

1. The columns in the wide format that we want to gather. Any number of columns can be gathered together using the *cols* argument by giving the names of the first and last columns in the column block we want to include e.g. *cols = female : male*.

2. The name we want to give to our variable column using the *names_to* argument.

3. The name we want to give to our value column using the *values_to* argument.

We can demonstrate the use of **pivot_longer()** using the `wide` data set from the example above to show how we can reverse the operation – moving from a wide format to a long format:

```
long <- wide %>%
   pivot_longer(cols = female : male,
                names_to = "gender",
                values_to = "count")
long
```

```
# A tibble: 30 x 4
# Groups:   course, village [15]
  course   village   gender count
  <chr>    <chr>     <chr>  <int>
1 forestry Anota     female     9
2 forestry Anota     male      17
3 forestry Lamaris   female     3
4 forestry Lamaris   male       4
5 forestry Maniwavie female     4
6 forestry Maniwavie male       8
# ... with 24 more rows
```

9.9 Ordering

Usually when we first explore our data, we want to examine observations associated with extreme values. Often we are doing this to check for mistakes or to see if the data has followed our expectations. The easiest way to do this is to order the data set by a variable that we are interested in.

9.9.1 arrange() – tidyverse(dplyr)

The **arrange()** function orders the data set in an ascending order based on the values in the column (top six rows shown):

```
hunt %>%
  arrange(age)

  hunter.ref age village distance hours catch.weight
1         H6  18 Takendu     2.37  5.47         8.72
2        H10  20   Anota     2.71  5.54         8.99
3         H4  22   Anota     4.02  4.21         8.62
4         H2  29 Takendu     2.13  6.29         8.29
5        H16  34 Lamaris     4.94  2.79         6.56
6        H17  34    Nugi     2.44  5.86         8.19
```

If we wanted a descending order we would nest the **desc()** function into the **arrange()** function e.g. `arrange(desc(age))`.

9.9.2 top_n() – tidyverse(dplyr)

Sometimes, when we are dealing with very large data sets we are only interested in seeing the highest or lowest ranked observations for a particular variable rather than seeing the whole data set. The **top_n()** function allows us to do this. However, we need to give **top_n()** the name of the column for the ranking, if none is given the function will automatically use the last column. The argument n defines the number of entries e.g. $n = 2$ is the two highest values. By using a negative value we can select the lowest values e.g. $n = -3$ would be the three lowest values. Note: the **top_n()** function does not automatically order – so you will may want to combine it with **arrange()** function. In the following example, we find the observations associated with the lowest three values of `catch.weight`:

```
hunt %>%
  top_n(n = -3, catch.weight) %>%
  arrange(catch.weight)

  hunter.ref age    village distance hours catch.weight
1          H5 60       Nugi     2.54  5.36         4.92
2         H13 57 Maniwavie     3.12  4.50         5.18
3          H8 52 Maniwavie     2.27  5.64         5.57
```

9.10   Joining

Often, the information we need for an analysis is not in one data frame but several. This can be because the data is stored across different files, or because we need to incorporate data from an outside source (such as a government data set).

In order to join two data sets together they must have a shared column known as a 'key' – which will allow matching between data sets. The key must be a character, not a factor. Provided these conditions are met the two data sets can be joined together in a variety of ways. However, if we were importing data frames using an older R version (< 4.0) we would need to use the **read.csv()** function with the *stringsAsFactors = FALSE* argument to ensure no columns were read as factors.

In order demonstrate different types of joins while avoiding issues with factors we will make two tibbles. As position is important in making joins we will call our tibbles **left** and **right** to make it obvious. The tibbles have information on 7 individuals – identified by a letter from 'A' to 'F' (**id**). First we will make a tibble called **left** which, along with its reference column (**id**), has a column of values called L. Next we will make a tibble called **right** which, along with its reference column (**id**), has a column of values called R:

```
L <- c(0, 1, 2, 3 )
id <- c("A", "B", "C", "D")
left <- tibble(id, L)

R <- c(0.1, 0.2, 0.3 , 0.4)
id <- c( "B", "C", "E", "F" )
right <- tibble(id, R)
```

There are four commonly used ways of joining data sets in the `tidyverse`:

1. **left_join()** – returns all rows from keys found in the left table, and the matched keys from the right table

2. **right_join()** – returns all rows from keys found in the right table, and the matched keys from the left table

3. **full_join()** – returns all rows from keys found in either data frame (the union of the two data frames)

4. **inner_join()** – only joins rows from keys found in both data frames (the intersection of the two data frames)

The code corresponding to these different situations is:

```
left_join(left, right, by = "id")
right_join(left, right, by = "id")
full_join(left, right, by = "id")
inner_join(left, right, by = "id")
```

The differences between these types of joins, and their effect on the final data set are shown in Figure 9.2.

9.11 Recommended resources

- Hadley Wickham and Garret Grolemund (2016) R for Data Science. O'Reilly Media, Sebastopol. A free online version is available at: https://r4ds.had.co.nz.

9.12 Summary

- The `tidyverse` package has a well developed system for data wrangling.

- The pipe %>% is a way of making a sequence of functions. This can make code easier to read.

- The **group_by()** and **summarise()** functions are particularly useful for generating tables required by project reports.

- There are two main types of data format: wide and long. It is important to know how to transform your data between the two types.

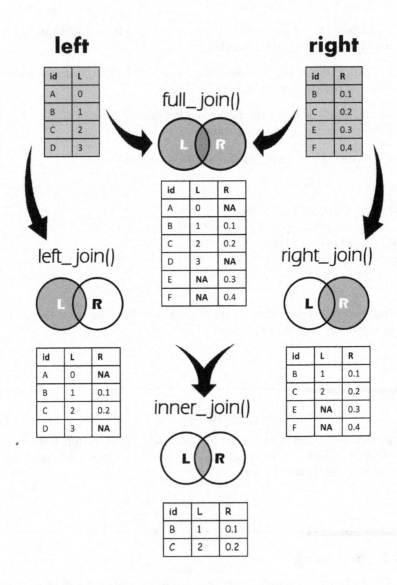

FIGURE 9.2
The relationships between the most common types of joins

10

Data cleaning

In this chapter we will look at data cleaning, focusing on:

- Fixing the names of columns.
- Fixing the names of factors.
- Replacing and removing missing values.
- Anonymising data.

10.1 Cleaning is more than correcting mistakes

This chapter is about data cleaning – getting data into a mistake-free form which is appropriate for analysis. If we are working with a data set which has been manually entered over many years, by many people, it is almost inevitable that there will be mistakes. Most data sets will have one or two issues (usually around inconsistent capitalisation and spelling, bad name choices, unhelpful symbols, and occasionally missing values). But, be prepared, occasionally you will encounter data sets which are unbelievably messy. Also, data cleaning doesn't just mean fixing common data entry mistakes, but also anonymising sensitive data. Conservation and development projects often such collect information and consequently there is a need to ensure this information is not accidentally shared or revealed during the reporting process.

10.2 The packages

For this chapter we will require the functions from:

```
library("janitor")
library("tidyverse")
```

10.3   The data

For this chapter we will use three example data sets (`clinic` , `messy`, and `split.data`) from the `condev` package. We'll begin by loading the `condev` package and the data sets:

```
library("condev")
data(clinic)
data(messy)
data(split.data)
```

Let's look at the `messy` data frame:

```
messy

    Village income_before INCOME..after. lang type   x
1    Anota           123            122  Vaina  NA
2     <NA>          $153            154  Vaina 101
3 maniwavie          177            173  Vaina  98
4     nugi           161            164 Binafa 102
5  Takendu           172            169 Binafa  98
6  Tamarua          <NA>              0   <NA> NaN
```

Well, this is a bit of a disaster – but not the worst I've seen. Clearly, this data frame has got a few problems:

1. The naming system of the column names is not consistent.

2. One of the column names (x) is not informative.

3. The factors in the `Village` column are capitalised differently.

4. The data frame has numerous `NA` results (indicating missing values[1]) as well as an `NaN` ('Not a Number' – a result you get when you divide by zero).

If our data frame has been produced via a spreadsheet and is small then it might be easiest just to fix the original spreadsheet (as it might take longer

---

[1]R uses `<NA>` rather than `NA` for factors and character columns to prevent the missing value being mistaken for a two letter entry called 'NA'

to write the R code to fix it than to correct it manually). However, if we don't want to alter the original data, or if there are too many entries to fix manually we can turn to R.

10.4 Changing names

The most common change we usually need to make is to fix the naming system used in the data set. Usually, we are renaming columns or changing factor names.

10.4.1 clean_names() - janitor

We can use the **clean_names()** function from the janitor package as a quick way to standardise column names. We could use the **rename()** function (described in the next section) but if our data frame has more than a few corrections then **clean_names()** function will save us a lot of time. This function is very simple to use and fixes column names automatically:

```
messy %>%
  clean_names()

    village income_before income_after lang_type   x
1     Anota           123          122    Vaina  NA
2      <NA>          $153          154    Vaina 101
3 maniwavie           177          173    Vaina  98
4      nugi           161          164   Binafa 102
5   Takendu           172          169   Binafa  98
6   Tamarua          <NA>            0     <NA> NaN
```

If there are any missing column names the **clean_names()** function will assign them a name begin with x (and thereafter x_2, x_3 etc.) The function also standardises capitalisation, spaces and punctuation using the default 'snake_case' form. If you prefer a different case five other options exist (run ?clean_names() to see options).

10.4.2 rename() - tidyverse(dplyr)

While **clean_names()** function can improve the appearance of column names it can't produce a more descriptive name. However, we can do this manually using the **rename()** function. It's basic form is:

```
rename(new.column.name = old.column.name)
```

In the `tidyverse` new names appears on the left hand side of the equals sign while the old names appears on the right hand-side. Ordinarily, R doesn't recognise multi-word names separated by spaces as variables (although data wrangling can result in these multi-word column names appearing). However, we can get R to recognise these names by enclosing them in back ticks (as in the case of the `lang type` column in the `messy` data set – see example below). If you do encounter multi-word variable names it is probably best to change them to a single word to make your coding simpler (alternatively, join the words together using a dot or underscore e.g. 'lang.type' or 'lang_type'):

```
messy %>%
    rename(income_after = INCOME..after.,
           language = `lang type`,
           percentage = x)
```

10.4.3 fct_recode() - tidyverse(forcats)

In order to change a factor name within a column we can use the **fct_recode()** function. Its basic form is:

```
fct_recode(column.name,
              "new.factor.name" = "old.factor.name")
```

The name of the column has to be included in the function, and the factor names have to be given in quotation marks. If there are any missing column names in the data set **fct_recode()** will return an error. If we wanted to capitalise the factors within the `Village` column of `messy` we would place the **fct_recode()** function within **mutate()**:

```
messy %>%
    mutate(Village = fct_recode(Village,
                                "Maniwavie" = "maniwavie",
                                "Nugi" = "nugi"))
```

10.4.4 str_replace_all() - tidyverse(stringr)

Sometimes people accidentally insert unnecessary characters during data entry that prevent R treating a column as numeric (e.g. $ and % signs), other times values might be too long to be practical (e.g. an animal given a tag number of 'R000000037980' could just be recorded as 'R37980'). In programming, a sequence of characters made of text, numbers, or symbols is known as a string. Ideally, if we need to alter strings in a data set we should import the data with an R version ≥ 4.0 [2], or alternatively if we are using an earlier R version: the **read.csv()** function with the argument *stringsAsFactors = TRUE*, or the **read_csv()** function. By doing this columns containing strings are read by R as characters (thereby avoiding issues associated with changing factor levels – see Chapter 13). We can then use the **str_replace_all()** function to replace or remove parts of the strings from our columns.

The **str_replace_all()** function needs three arguments: the variable we are altering, the string (characters) we want to replace, and the string we want to replace it with. Removing a string or character is just a case where the replacement string is empty. The basic form of the function is:

```
str_replace(column.name, "remove" , "replace")
```

Manipulating strings can be complex[3]. Here we will limit ourselves to the simplest situations. To demonstrate we will make a vector (**mixed**):

```
mixed <- c("AzaleaAnota", "Azalea?Anota", "Azalea?Anota!")
```

In the most simple situation we can remove a string of text characters:

```
str_replace_all(mixed, "Azalea", "")

[1] "Anota"   "?Anota"  "?Anota!"
```

If we want to remove a string which includes a special character (e.g. ?, %, $) we can do this by listing the character after two backslashes:

```
str_replace_all(mixed, "Azalea\\?", "")

[1] "AzaleaAnota" "Anota"        "Anota!"
```

---

[2]this is because R version ≥ 4.0 uses *stringsAsFactors = TRUE* as the default

[3]as a result there is a special language for describing patterns in strings known as regular expressions (or regexps for short)

If we want to replace a number of different characters regardless of position we can put these within square brackets " [ ] ". Any character, including special characters, within the square brackets will be replaced (no backslashes are required):

```
str_replace_all(mixed, "[Azalea?!]", "")

[1] "not" "not" "not"
```

We can now return to our **messy** data set and remove any $ found in the `income_before` column by using the **str_replace_all()** function within **mutate()**:

```
messy %>%
    mutate(income_before = str_replace(income_before,
                                "[$]" , ""))
```

10.5   Fixing missing values

The presence of missing values in R, as NAs or NaNs, will prevent many functions from producing meaningful results, for example:

```
mean(messy$x)

[1] NA
```

There are only two real options for missing values: replacing them or removing them. The choice depends on an understanding of the particular data set and how it is going to be used. If the data is required for calculations or modelling then every effort should be made to replace the missing data with real values (as most modelling in R automatically removes observations with missing values). Sometimes if those real values have been lost then it might be appropriate to estimate them (this is known as imputation and is not covered in this book). A final course of action would be to delete the individual rows of data that have the missing values. However, in some situations NAs and NaNs are perfectly acceptable, and indeed expected. For example, if we have made a table of training dates (e.g. for a donor report) we might expect an NA to appear if a particular training course did not occur on a particular date. Here the presence of an NA is not an issue, just that it is unsightly in the table. In such a case we might want to replace the NA with a better symbol (possibly a blank space, or a '–' sign).

In the following sections we look at some of the methods for handling replacing and removing missing values in the tidyverse and base R.

10.5.1 fct_explicit_na() - tidyverse(forcats)

In order to replace a missing value with a factor we need to use the fct_explicit_na() function. Its basic form is:

```
fct_explicit_na(column.name,
                "new.factor.name" = "old.factor.name")
```

In the example below we replace the NA in the Village column with a new factor called 'Lamaris' by using the fct_explicit_na() function within mutate():

```
messy %>%
  mutate(Village = fct_explicit_na(Village, "Lamaris"))
```

10.5.2 replace_na() - tidyverse(tidyr)

To replace a missing value in a column which is not a factor we use the replace_na() function. In this function we give a list (using the list() function) of the columns we want to change and the numbers or characters that we want to replace the missing values with. Its basic form is:

```
replace_na(list(column.1 = "new.value",
                column.2 = "new.value"))
```

In the example below we change the missing values in the x column to a '-' sign:

```
messy %>%
  replace_na(list(x = "-"))
```

10.5.3   replace() - base R

We can also replace continuous or character values through a logical rule using the **replace()** function with a logical operator (such as ==). This function will return a warning if run on a factor (and the factor will not be changed). The **replace()** function can be used to replace all values that correspond to a filtering rule. The basic form is:

```
replace(column.name,
        index.column == "filtering.rule", replacement.value)
```

In the example below we can replace all entries in column x with the value 99 when the Village is 'Anota':

```
messy %>%
    mutate(x = replace(x,   Village == "Anota", 99))
```

If we want to replace all the missing values with the same character or number we can use the **replace()** together with the **is.na()** function. The **is.na()** function identifies whether or not an NA or NaN is present and, if it is present, replaces it. The dot inside the **is.na()** function means the function will check all columns. Remember, this function cannot change the NA associated with a factor and consequently will give a warning that the factors present haven't been changed. Also, we don't have to place the **replace()** function within **mutate()** because we are altering the whole data set:

```
messy %>%
  replace(is.na(.), "-")
```

10.5.4   drop_na() - tidyverse(tidyr)

Sometimes, we will be left with no choice but to delete the observations associated with the missing values. The **drop_na()** function allows us to do this (and works on all data types including factors). If no column names are given all rows containing NA or NaN are removed. We can select which columns are affected through the basic form:

```
drop_na(column1)            # just column1 affected
drop_na(c(column1, column2)) # column1 and column2 affected
drop_na(column1:column67)   # column1 to column67 affected
drop_na(-column17)          # only column 17 not affected
```

The code below will drop all rows in `messy` which have an `NA` in the `lang type` column (notice that we have to use back ticks because it is a two-word name – see section 10.4.2).

```
messy %>%
  drop_na(`lang type`)
```

10.5.5 Cleaning a whole data set

We are now in a position to clean the `messy` data set. Because the data set is slowly changing each time a function is performed, the code differs slightly from the short examples we have shown in this chapter.

```
clean <- messy %>%
  clean_names() %>%
  rename(language = lang_type,
         percentage = x) %>%
  mutate(village = fct_recode(village,
                              "Maniwavie" = "maniwavie",
                              "Nugi" = "nugi"),
         village = fct_explicit_na(village, "Lamaris"),
         income_before = str_replace(income_before,
                                     "[$]" , ""),
         percentage = replace(percentage,
                              village == "Anota", 99))%>%
  drop_na(language)
```

The result is a clear, well-formatted data set:

```
clean

    village income_before income_after language percentage
1     Anota           123          122    Vaina         99
2   Lamaris           153          154    Vaina        101
3 Maniwavie           177          173    Vaina         98
4      Nugi           161          164   Binafa        102
5   Takendu           172          169   Binafa         98
```

10.6   Adding and dropping factor levels

Despite the neat appearance of the our `clean` data set (created in section 10.5.5) there are some hidden issues. If we examine the structure of the data frame we see that an unused factor level in `village` (`Tamarua`) hasn't been dropped:

```
levels(clean$village)

[1] "Anota"
[2] "Maniwavie"
[3] "Nugi"
[4] "Takendu"
[5] "Tamarua"
[6] "Lamaris"
```

Often, if we have just cleaned a data set we may want to drop any unnecessary levels. However, sometimes we will want to keep unused levels when we want to show that levels exist – even if there was no data associated with them. For example, if we were graphing the results of a multi-choice question we would want to show that 'question D' existed even if no one choose that option.

10.6.1   fct_drop() – tidyverse(forcats)

Often we will want to remove unused levels for simplicity. This can be done using the **fct_drop()** function. Its basic form is:

```
fct_drop(column.name, "dropped.factor.level")
```

For example, if we wanted to remove the 'Tamarua' level from `village` in `clean`:

```
dropped <- clean %>%
  mutate(village, village = fct_drop(village, "Tamarua"))

levels(dropped$village)
```

```
[1]  "Anota"
[2]  "Maniwavie"
[3]  "Nugi"
[4]  "Takendu"
[5]  "Lamaris"
```

10.6.2 fct_expand() - tidyverse(forcats)

We can also add factor levels using the **fct_expand()** function. We do this to include a factor which, for some reason, is not already present in a data set. An example would be when a multiple choice answer is absent in a data set because no respondent answered that way. Normally, in such a situation we would still want to indicate that an option for that answer existed. Its basic form is:

```
fct_expand(column.name, "new.factor.level")
```

For example, if we wanted to show that we could add a new level ('Sapul') to the village factor in clean:

```
expanded <- clean %>%
    mutate(village, village = fct_expand(village, "Sapul"))

levels(expanded$village)

[1]  "Anota"
[2]  "Maniwavie"
[3]  "Nugi"
[4]  "Takendu"
[5]  "Tamarua"
[6]  "Lamaris"
[7]  "Sapul"
```

10.6.3 Keeping empty levels in ggplot

In a ggplot missing levels are automatically dropped unless we add *drop = FALSE* to the **scale_***_discrete()**:

```
ggplot()+
  geom_point(data = clean, aes(x = language, y = village),
             size=4)+
  scale_y_discrete(drop = FALSE)+
  scale_x_discrete(drop = FALSE)
```

We can see the consequences of the different approaches to handling factor levels in Figure 10.1:

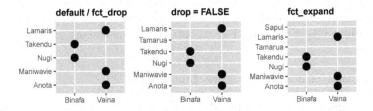

FIGURE 10.1
A graphic comparison of dropping and adding factors

10.7   Fusing duplicate columns

Sometimes when multiple people have worked on the same spreadsheet, they may accidentally duplicate columns but with different column names. Let's imagine a situation in which the data we need has been split across multiple columns. You can find this example in the `split.data` within the `condev` library:

```
split.data

    village income_before income_after language percentage before after
1     Anota            NA           NA     Vaina         99    123   122
2   Lamaris            NA           NA     Vaina        101    153   154
3 Maniwavie           177          173     Vaina         98     NA    NA
4      Nugi           161          164    Binafa        102     NA    NA
5   Takendu           172          169    Binafa         98     NA    NA
```

What we see is that all the score information is there but it is divided across differently named columns (i.e. `income_before` = before, and `income_after` = after) :

10.7.1 coalesce() - tidyverse(dplyr)

We can fuse any number of different columns together using the **coalesce()** function. The only rule is there must be a missing value in the receiving column. The order of the columns is important – the missing values of the receiving column will be substituted with values from the first non-NA donor column. Its basic form is:

```
coalesce(receiving.column, donor.columns...)
```

In this way we can fuse the receiving and donor columns of `split.data`. However, in order to clean the data set we will also want to remove any leftover donor columns. We can do this by using the using a negative sign with the **select()** function:

```
split.data %>%
   mutate(income_before = coalesce(income_before, before),
          income_after = coalesce(income_after, after)) %>%
   select(-c(before,after))
```

Which results in:

	village	income_before	income_after	language	percentage
1	Anota	123	122	Vaina	99
2	Lamaris	153	154	Vaina	101
3	Maniwavie	177	173	Vaina	98
4	Nugi	161	164	Binafa	102
5	Takendu	172	169	Binafa	98

10.8 Organising factor levels

If we have done a lot of manipulation with factors our factor levels are likely to be confused and not in the order we want for graphing (e.g. alphabetical). For example, `village` in our `clean` data set (created in section 10.5.5) currently has its levels in the following order:

```
levels(clean$village)

[1] "Anota"
[2] "Maniwavie"
[3] "Nugi"
[4] "Takendu"
[5] "Tamarua"
[6] "Lamaris"
```

10.8.1   fct_relevel() – tidyverse(forcats)

We can organise levels by using the **fct_relevel()** function. We can organise the levels by giving an explicit order. Its basic form is:

```
fct_relevel(column.name, "level.1", "level.2", "level.3")
```

We could also relevel the factors to be alphabetical by turning the factors into characters with the **as.character()** function and then sorting them using the **sort()** function:

```
fct_relevel(column.name, sort(as.character(column.name)))
```

Consequently, we can sort `village` to be alphabetical:

```
sorted <- clean %>%
    mutate(village =  fct_relevel(village,
                                  sort(as.character(village))))
```

This results in a change to the levels:

```
levels(sorted$village)

[1] "Anota"
[2] "Lamaris"
[3] "Maniwavie"
[4] "Nugi"
[5] "Takendu"
[6] "Tamarua"
```

10.8.2   fct_reorder() - tidyverse(forcats)

If we wanted to we could also organise levels by a variable using the **fct_reorder()** function. It takes the basic form:

```
fct_reorder(factor.column, variable.column)
```

For example, we can reorder `village` by `percentage`:

```
reordered <- clean %>%
  mutate(village = fct_reorder(village, percentage))
```

This results in a change to the levels (note: R will place the absent level ('Tamarua') at end of the list. This occurs regardless of whether the order is ascending or descending):

```
levels(reordered$village)

[1] "Maniwavie"
[2] "Takendu"
[3] "Anota"
[4] "Lamaris"
[5] "Nugi"
[6] "Tamarua"
```

10.8.3   fct_rev() - tidyverse(forcats)

Sometimes we may want to reverse the existing order of the factors. We can do this using the **fct_rev()** function. It takes the form:

```
fct_rev(factor.column)
```

As a result, we can reverse the original order of `village` levels found in `clean`:

```
reversed <- clean %>%
  mutate(village = fct_rev(village))
```

This results in a change to the levels:

```
levels(reversed$village)
```

```
[1] "Lamaris"
[2] "Tamarua"
[3] "Takendu"
[4] "Nugi"
[5] "Maniwavie"
[6] "Anota"
```

10.9   Anonymisation and pseudonymisation

Often data collected in conservation and development projects will contain information which is highly sensitive. Development data might contain sufficient information to allow individual people to be identified through: names, birth dates, addresses, government issued identity numbers, or from a combination of personal circumstances. Such data, if made public, is not only an invasion of privacy but can be exploited. Similarly, conservation data may inadvertently contain sufficient detail to allow exploitation (e.g. location data which allows vulnerable wildlife to be targeted by wildlife traders).

Data anonymisation is a large topic and cannot be covered here in any detail. But at its most basic we can understand that some data must be de-identified before it is used, studied, or published. The use of such data must follow the appropriate ethical protocols (which may be defined by the participating communities, your institution, or the government). When dealing with personal data the consent of the individuals involved and de-identification (removal of names) can be considered a minimum.

Data which has been altered to such an extent that it is not possible to match individuals against the original data can be considered fully anonymised. By comparison data which has only had its identifiers replaced with by pseudonyms, is often called pseudonymisation. As a result, pseudonymisation is weaker than anonymisation in terms of privacy. We will examine an example of pseudonymisation with the fictitious `clinic` data set. But before we begin we should look at its structure to identify any sensitive information:

```
str(clinic)
```

```
'data.frame': 368 obs. of 4 variables:
```

```
$ name : chr "Vivian Kodani" "Nancy Lim" "Celia Kearns" "Samin
Fifita" ...
$ village : chr "Maniwavie" "Maniwavie" "Anota" "Nugi" ...
$ visit.type: chr "birthdate" "birthdate" "birthdate"
"birthdate" ...
$ date : Date, format: "2018-09-04" ...
```

This data set has information on when women visited a health centre and the date they gave birth. Clearly, this data has some highly personal information. The `name` variable is clearly the most sensitive, but given the small size of the data set we would probably also want to de-identify `village`. We can accomplish much of this using the `tidyverse`. This is one of the situations when it is easier to work with factors than characters.

10.9.1   fct_anon() - tidyverse(forcats)

The **fct_anon()** function is specifically for anonymising data. The **fct_anon()** function replaces factor levels with random numeric identifiers. Neither the values nor the order of the levels of the original data set are preserved. If we want we can also add a series of characters to the new factor levels (called a prefix) through the *prefix* argument. The function takes the basic form:

```
fct_anon(column.name, prefix = "your.prefix")
```

Because the **fct_anon()** function works only with factors we need to check that the columns we want to anonymise are factors – if they are not factors we will need to use the **as.factor()** function to change them.

As outlined earlier, in order to fully anonymise the `clinic` data set we will not only need to anonymise the name of the individual but also their village, in order to prevent identification. As the `clinic` data set uses characters rather than factors in these variables we need to change them to factors first:

```
clinic %>%
  mutate(name = as.factor(name),
         name = fct_anon(name, prefix = "F"),
         village = as.factor(village),
         village = fct_anon(village, prefix = "V"))
```

Unlike other times we have used random number generation here we are deliberately not using **set.seed()** function – as we deliberately want to remove the ability to replicate the results. Consequently, your results will differ from the ones given below (top 6 rows of data are shown):

```
  name village visit.type         date
1  F56      V5  birthdate 2018-09-04
2  F20      V5  birthdate 2018-11-17
3  F29      V1  birthdate 2019-01-16
4  F41      V2  birthdate 2018-11-03
5  F50      V2  birthdate 2018-08-17
6  F08      V1  birthdate 2018-09-21
```

If for some reason, we wanted anonymise the patient's names but pseudonymise their villages (so villages might be identifiable to us but not others) we could use the **fct_recode()** function. So long as we keep the code we will have a record of the pseudonym for each village:

```
clinic %>%
  mutate(name = as.factor(name),
         name = fct_anon(name, prefix = "F"),
         village = as.factor(village),
         village = fct_recode(village,
                              "A" = "Maniwavie",
                              "B" = "Anota",
                              "C" = "Nugi",
                              "D" = "Takendu",
                              "E" = "Lamaris"))
```

Which in this instance, results in the following dataframe (first 6 rows shown) your results will differ:

```
  name village visit.type         date
1  F04       A  birthdate 2018-09-04
2  F31       A  birthdate 2018-11-17
3  F03       B  birthdate 2019-01-16
4  F32       C  birthdate 2018-11-03
5  F49       C  birthdate 2018-08-17
6  F18       B  birthdate 2018-09-21
```

10.10   Recommended resources

- Hadley Wickham and Garret Grolemund (2016) R for Data Science. O'Reilly Media, Sebastopol. A free online version is available at: `https://r4ds.had.co.nz`.

- `https://www.dataprotection.ie/en/guidance-landing/anonymisation-and-pseudonymisation`. Guidance on Anonymisation and pseudonymisation produced by the Data Protection Commission (Ireland) for individuals in the European Union.

10.11   Summary

- Very rarely will you work with a data set that is free of mistakes.

- Changing column names and fixing missing values is easy.

- Changing factors is more complicated because factors have ordered levels.

- If you have done a lot of factor manipulation the factor levels will out of sequence and may need to be fixed.

- Anonymisation is an increasingly important issue in data science. Both anonymisation and pseudonymisation can be undertaken with the `tidyverse`.

11

Working with dates and time

In this chapter we will look at:

- Getting R to recognise dates.

- Extracting date and time information.

- Working with time intervals.

- Graphing with dates.

11.1   The two questions

In conservation and development projects when we work with dates and times we usually have two types of questions that relate to time: (1) when did it happen, and (2) how long did it go for? These questions look like they should have simple answers but working with dates can be frustrating because:

- There are a large variety of date formats e.g. 10/11/19, 11-10-2019, 10Nov19, and 43779 are the same date in different formats.

- Times vary geographically thanks to time zones and use of daylight saving time in some countries.

- Times periods are inconsistent thanks to leap years and, occasionally, leap seconds.

Whether we are trying to transform wildlife tracking data obtained by satellites into local time, working out average response times for humanitarian emergencies, or just trying to sensibly graph dates or times – a few R tools, demonstrated in this chapter, can make our working lives much easier.

11.2   The packages

For this chapter we will require the functions from the following packages:

```
library("lubridate")
library("janitor")
library("tidyverse")
```

In particular we will be using the `lubridate` package for dealing with dates, but as `lubridate` is not part of the `tidyverse`, it will have to be loaded separately.

11.3 The data

In this chapter we will mainly use short code examples involving different date formats. This is because the formatting of differently structured dates is usually the main issue with dates and times. However, we will use the `clinic` data frame from the `condev` package later for the purpose of graphing dates.

```
library("condev")
data(clinic)
```

The use of personal data, especially medical histories, is highly sensitive. The use of such data must follow the appropriate ethical protocols, and at a minimum include the consent of the individuals involved and be appropriately anonymised so that individuals are not identifiable (see Chapter 10 for methods). The data of `clinic` example used here is entirely fictitious – neither the names nor events relate to real people.

11.4 Formatting dates

R is unable to automatically recognise dates. Because of this, we have to tell R the format of the date – the `lubridate` package makes this simple. However, we still need to understand how date notation works in base R as we will use this when graphing.

11.4.1 Formatting dates with `lubridate`

Formatting dates is easy with the `lubridate` as the package contains a number of functions for describing the different possible orders of the year (y), month (m), and day (d). Differences in punctuation are ignored:

```
ymd("2019-11-10")
dmy("10-11-19")
mdy("November 10th 2019")
mdy("11/10/19")
```

All of these will give the same answer of:

```
[1] "2019-11-10"
```

Similarly, we can use `lubridate` to recognise a time given in hours (h), minutes (m), and seconds (s). Again, differences in punctuation are ignored:

```
hms("20:11:59")

[1] "20H 11M 59S"

hms("20.11.59")

[1] "20H 11M 59S"
```

We can also combine all the elements to make a specific date and time:

```
ymd_hms("2019-11-10 20:11:59")

[1] "2019-11-10 20:11:59 UTC"
```

In this example you'll see that the R output is followed by the letters 'UTC' (an abbreviation for Coordinated Universal Time). UTC is the primary time standard for the world. As a result, it is the default time zone given to date-times in R. We will learn more about time zones in section 11.7.

11.4.2 Formatting dates with base R

It is important to understand how base R formats dates. Especially because the function we use to format dates in a ggplot, **scale_x_date()**, uses the base R notation. A summary of the most common codes are given in Table 11.1. These codes are always preceded by a % sign. Date formatting in base R uses the **as.Date()** function. Unlike `lubridate`, punctuation must be included:

```
as.Date("10/11/2019", format = "%d/%m/%Y")

[1] "2019-11-10"
```

TABLE 11.1
The common base R date codes

Code	Description	Example
%a	Abbreviated week day	Sun
%A	Full weekday	Sunday
%b	Abbreviated month	Nov
%B	Month name	November
%d	Day	10
%m	Month	11
%w	Day of week (Sun = 0)	0
%y	Year without century	19
%Y	Year with century	2019

11.4.3 Numerical dates

Sometimes we will have a data set which has the date recorded as a numeric value (as a result of a software specific date counting system). So long as we know the date from which the computer system started counting this isn't a problem. We can use the **as.Date()** function from base R to make these into a date using the *origin* argument:

```
as.Date(43779, origin = "1899-12-30") # Windows Excel

[1] "2019-11-10"

as.Date(42317, origin = "1904-01-01") # Mac pre-2011

[1] "2019-11-10"
```

If we are working with data made in Excel then the `janitor` package has the **excel_numeric_to_date()** function to convert excel numeric values to dates. The advantage with this function is we don't have to know the date of the origin:

```
excel_numeric_to_date(43779, date_system = "modern")

[1] "2019-11-10"

excel_numeric_to_date(42317, date_system = "mac pre-2011")

[1] "2019-11-10"
```

11.5 Extracting dates

If we are given given a date-time object, for example:

```
x <- ymd_hms("2019-11-10 20:11:59")
```

We can extract different date and time components using the `lubridate` functions below. However, these outputs will no longer be dates – but numeric (or ordered) factors:

```
year(x)
```

```
[1] 2019

 month(x)

[1] 11

 day(x)
      \
[1] 10

 hour(x)

[1] 20

 minute(x)

[1] 11

 second(x)

[1] 59
```

We can also extract the days and names of the week:

```
 wday(x)

[1] 1

 wday(x, label = TRUE)

[1] Sun
Levels: Sun < Mon < Tue < Wed < Thu < Fri < Sat
```

In the same way we can get the name of the month by:

```
 month(x, label = TRUE)

[1] Nov
12 Levels: Jan < Feb < Mar < Apr < May < Jun < Jul < ... < Dec
```

11.6   Time intervals

If we want to know the duration of a time interval we can simply work out the time difference by subtraction of the two objects:

```
start <- ymd_hm("2017-09-27 08:11")
finish <- ymd_hm("2017-10-1 20:12")
finish - start

Time difference of 4.500694 days
```

Sometimes, the default units – such as in this example – are not very particularly helpful. Often it is better to use the **as.period()** function from lubridate – this way we get an output with easy to read units:

```
as.period(finish - start)

[1] "4d 12H 1M 0S"
```

Or if we prefer a numeric version, in a specific single unit (e.g. hours), it is easiest to use the **as.numeric()** function from base R:

```
as.numeric(finish - start, "hours")

[1] 108.0167
```

11.7   Time zones

If we are getting very specific about our times (i.e. using a date and time combination) then we probably want to specify the time zone too. The codes for the time zones normally take the form 'Region/City'[1]. A list of time zones can be found at:

https://en.wikipedia.org/wiki/List_of_tz_database_time_zones

---

[1]sometimes also given as 'Region/Sub-region/City' or 'Region/Sub-region'

The advantage of using these codes is that we can account for local changes in time (e.g. changes due to daylight saving time or assignment of time zones). We can specify a time zone directly using the *tz* argument when we create the value:

```
ymd_hms("2017-09-30 20:11:59", tz = "Pacific/Port_Moresby")

[1] "2017-09-30 20:11:59 +10"
```

But more often we will have a date-time variable lacking a time zone. To force a time zone we can use the **tz()** function:

```
my.time <- ymd_hm("2017-09-30 20:11")
tz(my.time) <- "Asia/Gaza"
my.time

[1] "2017-09-30 20:11:00 EEST"
```

11.7.1 The importance of time zones

Some countries and/or their sub-regions operate on daylight saving time which means local clocks shift by an hour in these regions twice a year. The timing of the change is entirely arbitrary. Some of these places have used, abandoned, and then resumed the practice of daylight saving time through the years. Additionally, some countries have changed their time zones. Such changes can affect the calculation of time durations. Fortunately, the *tz* argument within a date built by lubridate automatically accounts for these changes.

Consider an example in which we are recording the response time of emergency services. If the ambulance left at 01:55 am and arrived at 03:03 am what was the response time? The answer depends on where they were. If they were in Fiji on 4 November 2018 the response time was only 8 minutes despite the local time showing a change of over an hour. This is due to daylight saving time beginning at 2 am that day:

```
departed <- ymd_hm("2018-11-4 01:55", tz = "Pacific/Fiji")
arrived <- ymd_hm("2018-11-4 03:03", tz = "Pacific/Fiji")
as.period(arrived - departed, "hours")

[1] "8M 0S"
```

For this reason if we are recording durations it is sometimes important to include the time zone in the data. To check a time zone we can use the **tz()** function:

```
tz(departed)

[1] "Pacific/Fiji"
```

However, the time zone connected with a normal date-time will not be returned in the 'Region/City' format but rather as a difference in hours from UTC or as a time zone abbreviation[2]. In the example below, our time zone (Pacific/Fiji) is ahead of UTC by 12 hours:

```
departed

[1] "2018-11-04 01:55:00 +12"
```

11.7.2 Same times in different time zones

We can easily work out the time an event took place in a different time zone using the **with_tz()** function. For example, the eruption of Mt Pinatubo occurred at 1:42 pm on 15 June 1991 in the Philippines[3]:

```
Pinatubo <- ymd_hm("1991-6-15 13:42",
                   tz = "Asia/Manila")
```

Would have occurred at:

```
with_tz(Pinatubo, tzone = "Pacific/Honolulu")

[1] "1991-06-14 19:42:00 HST"

with_tz(Pinatubo, tzone = "Asia/Jakarta")
```

---

[2]see https://en.wikipedia.org/wiki/List_of_time_zone_abbreviations
[3]a massive eruption which killed more than 700 people and temporarily cooled the Earth by 0.5°C

```
[1] "1991-06-15 12:42:00 WIB"

with_tz(Pinatubo, tzone = "Africa/Lagos")

[1] "1991-06-15 06:42:00 WAT"
```

You'll notice that the **with_tz()** function returns the date-time with the time zone abbreviation. The 'Region/City' code has not been lost – it still can be found using the **tz()** function.

11.8   Replacing missing date components

On some occasions we may have a variable which is not a true date e.g. often months and years are recorded but not the actual date. Such missing values will prevent the variable being recorded as a date format in R and therefore prevent easy graphing. This situation can be corrected using the **parse_date_time()** function by giving the order of the date-time using the letter codes found in Table 11.1 (without the % sign). Take for example if a variable is just recorded as the month and year (e.g. "Mar1971"):

```
parse_date_time("Mar1971", orders = "mY")
```

This then assigns the first day of the month to the missing day component:

```
[1] "1971-03-01 UTC"
```

11.9   Graphing: a worked example

We can graph dates and times easily as ggplot can handle dates as axes. In the following example we can examine the relationship between the dates on which the women visited a health clinic (using the `clinic` data from the `condev` package).

A simple ggplot (with the names removed) would look like:

```
ggplot()+
  geom_point(data = clinic, aes(x = date, y = name))+
  theme(axis.text.y = element_blank())
```

Which would produce a messy graph looking like:

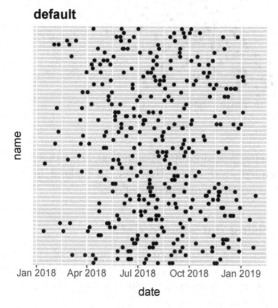

At first glance it appears the data is rather random. However, as we shall see, there is a pattern but because the y-axis is not in any kind of order we cannot see it.

11.9.1 Reordering a variable by a date

Often reordering the y-axis by the date can reveal a pattern:

```
ggplot()+
  geom_point(data = clinic, aes(x = date,
                                y = reorder(name, date)))+
  theme(axis.text.y = element_blank())
```

reordering by date

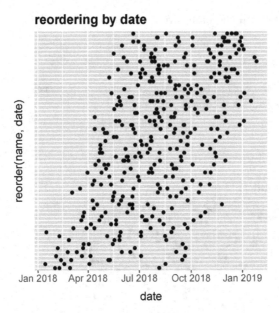

Looking at the density of dots it appears that the women who attended the clinic later attended more often. However, to make a more informative graph we will need to create some summary data.

11.9.2 Summarising date-based data

We can summarise our data with regard to the number of visits each woman had, when their first and last visits were, and the proportion of their pregnancy that was observed by the clinic (using 280 days as the typical pregnancy length). The following functions will help us:

  - **group_by()**: for a summary of each named individual.
  - **n_distinct()**: for the number of distinct dates.
  - **min()**: for the first date for each person.
  - **max()**: for the last date for each person.
  - **as.numeric()**: using the argument *"days"* for the time period in days.

We can use these functions to produce the following code:

```
time <- clinic %>%
  group_by(name) %>%
```

```
summarise(visits = n_distinct(date),
          first = min(date),
          last = max(date),
          proportion = as.numeric(max(date) - min(date),
                                   "days") / 280)
time
```

Which results in the following tibble (first 6 rows shown):

```
# A tibble: 60 x 5
  name                   visits first      last       proportion
  <chr>                  <int> <date>      <date>          <dbl>
1 Alexandra Nguyen           7 2018-04-13 2018-11-18      0.785
2 Alimansar Suzuki           7 2018-04-17 2018-11-25      0.794
3 Alisa Nolte                6 2018-03-06 2018-09-24      0.724
4 Angela Makam               8 2018-05-26 2018-12-28      0.774
5 Ann Patrick-Curley         7 2018-06-04 2019-01-22      0.832
6 Anna Tieu                  7 2018-06-24 2018-11-30      0.569
# ... with 54 more rows
```

By using graphing the number of visits against first visitation date, and overlaying the **geom_smooth()** function using a linear trend line (*method* =*"lm"*) we can see there it appears that women attended the clinic more often as time progressed:

```
ggplot()+
  geom_point(data = time, aes(x = first, y = visits),
             size = 5, alpha = 0.3)+
  geom_smooth(data = time, aes(x = first, y = visits),
              method = "lm",  alpha = 0.5, colour = "black")+
  labs(title = "overlay",
       x = "Date of first visit",
       y = "Number of visits")
```

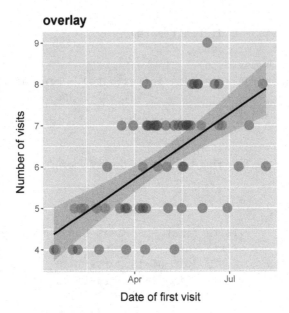

11.9.3   Date labels with scale_x_date() - tidyverse(ggplot2)

However, in the graph above we can see that the x-axis labels are too sparse. We can change this by using the **scale_x_date()** function. The function takes the form:

```
scale_x_date(date_labels = "%date.codes",
             date_breaks = "number period")
```

If we wanted to show the abbreviated month and year on the x-axis with tick marks every month we could use:

```
ggplot()+
  geom_point(data = time, aes(x = first, y = visits),
             size = 5, alpha = 0.3)+
  geom_smooth(data = time, aes(x = first, y = visits),
              method = "lm",  alpha = 0.5, colour = "black")+
  labs(title = "overlay with scale_x_date",
       x = "Date of first visit",
       y = "Number of visits")+
  scale_x_date(date_labels = "%b %y",
               date_breaks = "1 months")
```

Which will result in a much more readable graph:

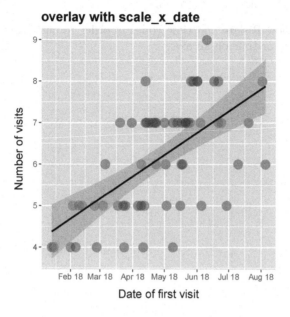

11.10 Recommended resources

- Hadley Wickham and Garret Grolemund (2016) R for Data Science. O'Reilly Media, Sebastopol. A free online version is available at: https://r4ds.had.co.nz.

- https://lubridate.tidyverse.org/. The official lubridate web page within the tidyverse. It includes a very useful lubridate cheatsheet.

11.11 Summary

- We have to tell R to recognise a set of characters as a date.

- When we extract date and time features these will no longer be recognised as dates.

- We can easily convert time intervals into different units.

- Time zones are not only important for working between regions but also for accurately measuring time durations within time zones (especially for countries using daylight saving time).

- We need to know R date and time codes in order to customise a ggplot which uses date or time on an axis.

12

Working with spatial data

In this chapter we introduce:

- Working with spatial data in R.

- Making maps.

- Manipulating maps.

- Extracting samples and other information from spatial data.

12.1 The importance of maps

One of the most important pieces of information required in any project report is a map of the project area or one showing where the data was gathered. A map not only helps record the position of a project or study but puts that information into context, for example, in relation to political borders, other human settlements, landscape features, or ecosystems.

While there are plenty of sophisticated mapping programs they are often expensive and time consuming to learn. However, we can easily build upon on our existing skills to make high quality maps in R which are both informative and rewarding to make. In addition to this we can use R to extract critical pieces of information from pre-existing map data.

It is worth pointing out that R is increasingly being used for spatial data and as a consequence there are a large amount of packages and specialised books for this topic. This chapter only covers the most basic approaches.

12.2 The packages

Spatial data is a fast developing area with many different R packages available. In this chapter we will focus on two key packages for dealing with each of the data types:

1. `sf` package: for vector data.

2. `raster` package: for raster data.

In order for the `sf` package to work we will need to download the Geospatial Data Abstraction Library first (GDAL). This is an external library (not an R library) which will allow us to work with different kinds of spatial data formats. Below are my current recommendations for downloading GDAL[1]:

- For Linux: run `apt-get install python-gdal` in the terminal.

- For Mac: go to `https://www.kyngchaos.com/software/frameworks/`.

- For Windows: go to `https://trac.osgeo.org/osgeo4w/`[2].

High quality maps can be made using the `tidyverse` package in combination with the `sf` (simple features) package. In this chapter we will also use a couple of helper packages: `units` for changing spatial units, `ggspatial` for scalebars and north arrows, and `rnaturalearth` for accessing world map data. Also we will use the `raster` package to explore some of R's real data power.

```
library("ggspatial")
library("raster")
library("rnaturalearth")
library("sf")
library("tidyverse")
library("units")
```

R may also ask you to install the following packages to allow the current packages to function properly (if not already installed):

```
install.packages(c("lwgeom", "sp", "rgeos", "rnaturalearthdata"))
```

---

[1]sometimes installing GDAL can be a little tricky and you may have to search the internet for a solution that suits your current computer setup

[2]this will also give you the option of loading QGIS which is a useful mapping alternative to the highly expensive ARCGIS software, however, QGIS is very large and will consume a large amount of data

12.3 The data

For this chapter we will use a number of example data sets from the `condev` package. Start by loading `condev` and the required data sets:

```
library("condev")
data(location)
data(treecover)
data(waihi)
```

12.4 What is spatial data?

Spatial data is data which represents a location in space (from which we get the name 'spatial'). In conservation and development that space is usually a geographic location on earth. There are two types of spatial data (see Figure 12.1):

1. Vector data: which can be shown as points, lines or polygons. These can be used to represent real world features e.g. towns (points), roads (lines), and national borders (polygons). Additionally, joining points makes a line, and enclosing an area with a line creates a polygon.

2. Raster data: which can be shown as a grid of pixels (e.g. a digital photo). Within a raster has each pixel has a value. Common raster types include elevation data, and land-use data.

12.5 Introduction to the sf package

The `sf` package stands for 'simple features', and is a relatively new way of handling spatial data in R. This package simplifies spatial data types by handling them as data frames. The `sf` package will read your vector data in the same way regardless if it is a point, line or polygon. As a result, it is very easy to use `sf` in combination with the `tidyverse` to turn ggplots into high quality maps. In the code below you will see the prefix 'st_' this stands for 'spatial type' which refers to any spatial data which is not an `sf` type.

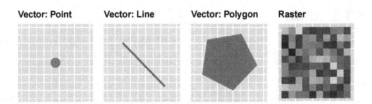

FIGURE 12.1
Examples of the basic spatial data types

12.5.1   Reading data: st_read() - sf

The **st_read()** function can read most types of spatial data. It does by talking to GDAL which you need to have installed (see the data section in this chapter for instructions). The **st_read()** function reads in the named file which has to include its file extension. For example to read a file made using keyhole markup language (which has a file extension '.kml'):

```
my.file <- st_read("my_file_name.kml")
```

The easiest way to ensure R can find your spatial data is to have it in the same folder as the project you are working on (by using 'projects' in RStudio (see Chapter 4) – this way you don't need to name the path every time). Alternatively, you can find find the file path using the instructions in Chapter 5.

12.5.2   Converting data: st_as_sf() - sf

Spatial objects created by package **sp** (which was until recently the standard way of handling spatial data in R) can be formatted to a simple features format using the **st_as_sf()** function:

```
waihi.sf <- st_as_sf(waihi)
```

12.5.3   Polygon area: st_area() - sf

Most conservation and development proposals require a measure of the 'project footprint' – the area where the project intervention is expected to have an effect. In order to make such an estimate we need to be able to measure the area of a polygon. We can easily get such an area using the **st_area()** function which will give the area in m^2:

```
waihi.area <- st_area(waihi.sf)        # area in square m
waihi.area

530993468 [m^2]
```

The units of this output can be changed to using the **set_units()** function from the **units** package in which we give the units we require (e.g. km², ha etc.):

```
set_units(waihi.area, km^2)        # area in square km

530.9935 [km^2]
```

Unfortunately, ggplot doesn't known how to graph 'units'. So we will have to modify any output by the **as.numeric()** function to return a graphable value.

12.5.4   Plotting maps: geom_sf() - tidyverse(ggplot2)

We can map the **sf** data using the **geom_sf()** function. Unlike most ggplot geoms it doesn't require x and y arguments, as it automatically reads these from the geometry feature of an sf object (which is contained within a column called 'geometry'):

```
ggplot() +
  geom_sf(data = waihi.sf, fill = "grey")
```

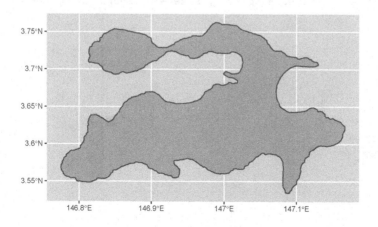

12.5.5 Extracting coordinates st_coordinates() - sf

Sometimes we will want to extract the actual coordinates from the **sf** object and turn them into a separate data frame. We might choose to do this when we want the coordinates but not the accompanying data (e.g. section 12.8.2). We can do this through the use of the **st_coordinates()** function:

```
waihi.xy <- st_coordinates(waihi.sf) %>%
  as.data.frame()
```

12.6 Plotting a world map with rnaturalearth

The **rnaturalearth** package provides a way to download and access pre-downloaded spatial data from https://www.naturalearthdata.com/. To access the pre-downloaded data country polygons we use the **ne_countries()** function making sure we are using the argument *returnclass = "sf"*.

```
world <- ne_countries(scale = "medium", returnclass = "sf")
```

We can still overlay other geoms with **geom_sf()** in the normal manner. But we have to make sure these geoms use the $x=X$ and $y=Y$ aesthetics to match the **sf** data, for example:

```
ggplot() +
    geom_sf(data = world, fill = "grey", colour = NA)+
    geom_point(data = waihi.xy, aes(x = X, y = Y),
               colour = "black", size = 5, shape = "square")
```

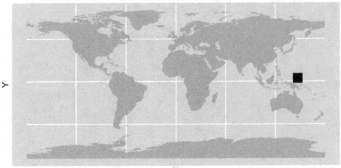

12.6.1 Filtering with filter() - `tidyverse(dplyr)`

Many conservation and development projects occur at regional levels or continental levels. Fortunately, **sf** data, being a data frame can easily be filtered. This means that we can use the **filter()** function to subset the map data to just the regions we are interested in, for example if we just want to consider the continent of Africa:

```
africa <- world %>% filter(continent == "Africa")

ggplot()+
    geom_sf(data = africa, fill = "grey")
```

Will produce:

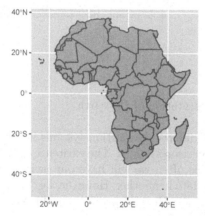

Additionally, just like a normal data frame we can get an overview of any variables associated with **africa** using **head()**, or, if we are using RStudio, **View()**.

12.7 Coordinate reference systems

The planet we live on is a sphere. But unfortunately spheres are not very practical for display purposes in reports. As a consequence, we have to transform the 3-dimensional earth into a 2-dimensional form known as a map.

The way we project (i.e. flatten) the earth onto a 2-dimension surface makes a big difference to how some areas look. We can see this in Figure 12.2. When we compare the image of the earth and the map we see the areas near the poles have become stretched. We can prove this by comparing grid lines (known as graticules) on earth to those on the map. What we can see is that while the graticules on the map appear to be parallel, the true distance between graticules is actually decreasing as we get closer to the poles.

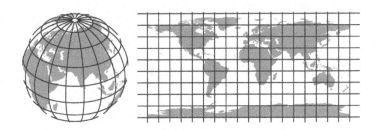

FIGURE 12.2
Spherical versus flat representation of the earth

There are hundreds of different projection systems in use today including many country specific projection systems. The existence of all these projection systems can cause issues as often we will have our base map in one kind of projection system but our data in another. However, we can transform between projection systems if we know their respective coordinate reference systems (CRS for short).

The two commonly used forms of CRS are an epsg code[3] (see `https://epsg.io/` for a searchable database) and something known as a PROJ.4 string (which can include information on the projection type, any regional zones, the datum (which outlines the map's coordinate system and reference points), and an ellipsoid which approximates the earth's roundness).

We can see the effect of different CRS when we map Greenland. First let's consider the standard longitude and latitude system (epsg = 4326) which is great for mapping on a 3-dimensional sphere but not so good for mapping onto a 2-dimensional map (as it distorts polar areas). To make this map we set the CRS code by putting it into the **st_crs()** function. We can then embed this code within the **coord_sf()** function so the maps will be plotted with the new projection :

---

[3]initially developed by the European Petroleum Survey Group

```
greenland <- world %>% filter(name == "Greenland")

ggplot() +
    geom_sf(data = greenland, fill = "black")+
    coord_sf(crs = st_crs(4326))
```

Now let's compare a Universal Transverse Mercator (UTM) projection which is centered on Greenland (crs = 32623, equivalent to UTM Zone 23 in the northern hemisphere) to the longlat projection. The UTM projection has much less distortion and better matches the actual area of the land mass. However, the downside of UTM projections is they distort more as we move away from the centre of the CRS:

```
ggplot() +
    geom_sf(data = greenland, fill = "black")+
    coord_sf(crs = st_crs(32623))
```

Below is the comparison of the two maps created by our code – notice how distorted the longlat version becomes when flattened:

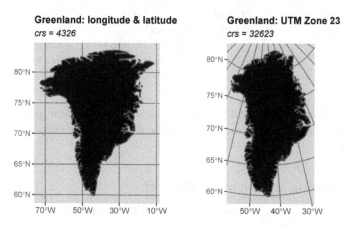

12.7.1 Finding the CRS of an object with st_crs() - sf

Details of the CRS of a spatial object can be found using the **st_crs()** function and examining the object's **$epsg** and **$proj4string** values:

```
our.map <- st_crs(greenland)

our.map$epsg

[1] 4326

our.map$proj4string

[1] "+proj=longlat +datum=WGS84 +no_defs"
```

We can see from the espg code it is 4326, and the proj4string tells us it is a longlat projection using the World Geodetic System 1984 (WGS84) as its datum.

12.7.2 Transform the CRS with st_transform() - sf

To permanently change the projection of a **sf** object we can use the **st_transform()** function by either using the epsg code or the projection information. Importantly, as the projection information is a string if we use different spacing an error will result.

```
greenland.trans1 <- st_transform(greenland, crs = 32623)
greenland.trans2 <- st_transform(greenland,
                              crs = "+proj=utm +zone=23
                              +datum=WGS84 +no_defs")
```

As both of these are equivalent they will result in the same projection, which we can check by comparing the epsg and proj4string outputs of the **st_crs()** function.

12.7.3 Cropping with coord_sf() - sf

If we want to crop the extent (the square edged boundary) of the map for a ggplot we can do this by giving the minimum and maximum values of the x-axis and y-axis to the **coord_sf()** function using the *xlim* and *ylim* arguments. For example, we can map a subsection of western Africa:

```
ggplot() +
    geom_sf(data = africa, fill = "grey")+
    coord_sf(xlim = c(-19, 13), ylim = c(4, 20))
```

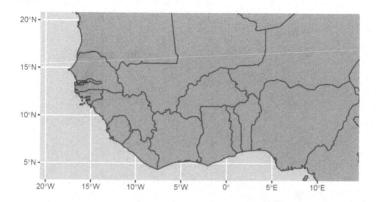

Traditionally, latitude and longitude coordinates are tagged with a letter to represent the hemisphere they are referring to (i.e. N, S, W, E). In order to express these hemispheres as numbers – northern and eastern hemispheres are given positive values, while southern and western values are given negative values.

12.8 Adding reference information

Up to now we haven't really added any information to our map – all we have been doing is graphing a polygon. Only by adding other reference information do we begin to turn it into a map.

12.8.1 Adding a scale bar and north arrow - ggspatial

By using the **annotation_scale()** and **annotation_north_arrow()** functions from the ggspatial package we can add a scale bar and north arrow. The positioning system of these features is controlled by the *location* argument (e.g. *"tr"* = top right, *"bl"* = bottom left etc). By using the *pad_x* and *pad_y* arguments we can make minor adjustments to the positioning:

```
ggplot() +
    geom_sf(data = waihi.sf, fill = "grey") +
```

```
annotation_scale(location = "tr")+
annotation_north_arrow(location = "tr",
                       pad_x = unit(0.1, "cm"),
                       pad_y = unit(1, "cm"))
```

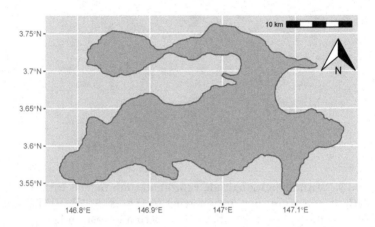

12.8.2   Positioning names with centroids - sf

Centroids can be thought of as the geometric centre of a polygon. As a consequence centroids can be useful for positioning the location names[4]. The **st_centroid()** function calculates the position of the centroids and **st_coordinates()** function returns the location of the centroids (as two columns called X and Y) which we can then bind to our `africa` data frame using the **cbind()** function.

```
africa.with.names <- st_centroid(africa$geometry) %>%
  st_coordinates() %>%
  cbind(africa)
```

We can then add these names to the map using the **geom_text()** function:

---

[4]using centroids for longlat coordinates across larger areas can be inaccurate, and as a consequence R will give a warning

```
ggplot() +
    geom_sf(data = africa, fill = "grey", colour = "white")+
    geom_text(data = africa.with.names,
              aes(x = X, y = Y, label = name),
              color = "black", fontface = "bold")+
    coord_sf(xlim = c(-19, 13), ylim = c(4, 20))
```

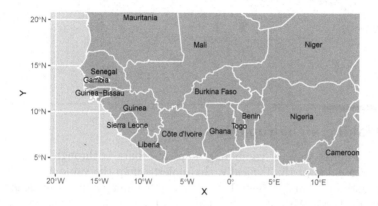

12.8.3   Adding names with geom_text() - `tidyverse(ggplot)`

We can also name locations to a map in a ggplot by using the **geom_text()** function combined with either a data frame, or providing x and y coordinates. If we are combining these names with points we need to include an adjustment with the argument *vjust* (vertical adjustment) or *hjust* (horizontal adjustment) to prevent the name being written over the top of the point:

```
ggplot() +
  geom_sf(data = waihi.sf, fill = "grey", colour = NA)+
  geom_point(data = location, aes(x = long, y = lat), size=3)+
  geom_text(data = location, aes(x = long, y = lat,
                                 label = village),
            vjust = 2, size = 3)+
  geom_text(aes(x = 146.98,y = 3.625, label = "Waihi Island"),
            fontface = 3, colour = "white", size = 6)
```

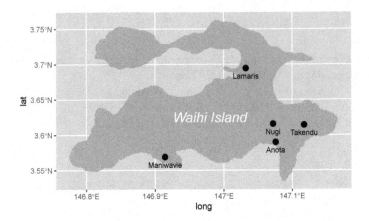

12.9 Making a chloropleth

One of the simplest ways of combining maps and data is through a visualisation called a chloropleth. Typically, chloropleths map the value of variable through a shade or colour gradient. Let's make some human population density data to demonstrate. We will use the **st_area()** function in combination with **set_units()** and **as.numeric()** functions to make some area data. From this area data we can calculate density per km^2 using the variable `africa$pop_est` contained within our data set:

```
africa$area.km2 <- st_area(africa) %>%
  set_units(km^2) %>%
  as.numeric()

africa$density.per.km2 <- africa$pop_est / africa$area.km2
```

To finish the chloropleth we then have to modify the shading of the map by using the **scale_fill_gradient()** function. Additionally, we can also make an object to control the breaks that will appear in the legend. As population density in Africa ranges more than 100 × between countries, a log scale is appropriate (otherwise it is hard to distinguish between countries based on colour alone). As a consequence, we need to use the *trans = "log"* argument. Often some regions have no data (e.g in our example – Western Sahara). In such situations it is important to represent the absent data by using a colour different to the value scale. This is so that an absence of data is not confused with actual data (e.g. a real value of zero or some other low value). This can be done by naming the colour using the *na.value* argument:

```
my_breaks = c(2, 5, 10, 20, 50, 100, 200, 400)

ggplot()+
    geom_sf(data = africa, aes(fill = density.per.km2))+
    scale_fill_gradient(low = "grey90", high = "black",
                        na.value = "white",
                        name = "density per km2",
                        trans = "log",
                        breaks = my_breaks,
                        labels = my_breaks)
```

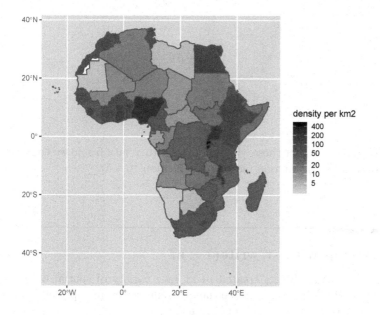

12.10 Random sampling

An important part of most survey designs is the need for random sampling (see Chapter 14). As spatial areas usually have irregular boundaries it is best to use a specialised function such as **st_sample()** to generate the samples (as

it only samples within the polygon area). In the example below we generate
20 random sampling locations:

```
set.seed(17)
sample.points <- st_sample(waihi.sf, 20)

ggplot() +
    geom_sf(data = waihi.sf, fill = "grey", colour = "white")+
    geom_sf(data = sample.points, shape = 3, size = 4)
```

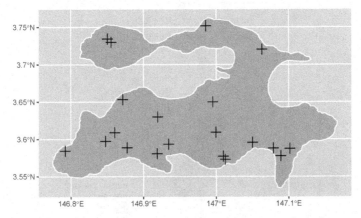

If we needed to extract these sample points into a data frame we could use
the **st_coordinates()** function (see code in section 12.5.5).

12.11   Saving with st_write() - sf

We can also save our objects created in **sf** in a variety of spatial file types
using the **st_write()** function. To see the different spatial file options (known
as drivers) we can use the **st_driver()** function:

```
st_drivers()
```

For example, if we wanted to save our randomly sample points on to a
GPS device we would reference the object we want to save, the name of the
new file, and use *driver = "GPX"*:

```
st_write(sample.points, "samplepoints.gpx", driver = "GPX")
```

12.12 Rasters with the raster package

Thanks to satellite technology and the internet, people can now access large data sets stored as rasters[5]. Such data sets are particularly useful for landscape scale conservation and land-use planning (e.g. identifying habitats, prioritising areas to protect, and establishing where conflicts over resources may occur). Raster files are often huge and can contain millions of data points as pixels. While raster data can be converted to a data frame it is usually far faster to do any calculations and other manipulations within the **raster** package. For the purposes of this chapter we only consider the simplest uses of the **raster** package involving a raster with a single layer.

12.12.1 Loading rasters

The **raster** package can read most types of gridded data files. One of the most common type of rasters data is the GeoTIFF format (with the file extension .tif or .tiff). The **raster()** function reads in the named file – which has to include the file extension. For example:

```
my.raster <- raster("global_forest_cover.tif")
```

12.12.2 Raster data

To examine a raster held in the **condev** package we simply call the object (e.g. **treecover**) in the same manner as a normal file:

```
treecover
```

This will result in a summary of its features:

---

[5]such as the Global Forest Cover data set: https://earthenginepartners.appspot.com/

```
class : RasterLayer
dimensions : 828, 1800, 1490400 (nrow, ncol, ncell)
resolution : 0.0002777778, 0.0002777778 (x, y)
extent : 146.7099, 147.2099, 3.530861, 3.760861 (xmin, xmax,
ymin, ymax)
coord. ref. : +proj=longlat +datum=WGS84 +no_defs +ellps=WGS84
+towgs84=0,0,0
data source : in memory
names : waihi_treecover
values : 0, 100 (min, max)
```

In the case of (e.g. **treecover**): its features are:

- **class:** the type of raster it is.

- **dimensions**: the number of columns and rows making up the raster's grid. and the total number of cells. In this example there are almost 1.5 million cells.

- **resolution:** the size of the pixels in terms of the coordinate system (i.e. in longitude and latitude it is 0.00028 of a degree)

- **extent:** the borders of the grid in the coordinate system.

- **coord. ref.:** the projection type.

- **data source:** origin of the data.

- **names:** the names of the raster layers.

- **values:** the minimum and maximum values of the raster pixels.

12.12.3   Plotting rasters

Plotting rasters is straight forward using the **plot()** function:

```
plot(treecover)
```

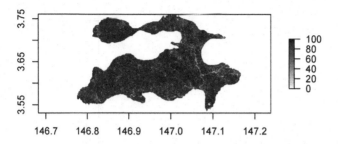

Despite the large number of data points we can quickly subset the data within a plot using standard operators. For example, we might only want to plot areas of our raster where `treecover` was greater than 80 (%):

```
plot(treecover > 80)
```

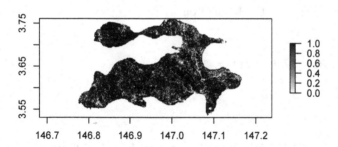

We can also examine the distribution (using a histogram) of the values in the raster layer using the **hist()** function from the **raster** package[6]. By default this histogram will produce a graph with an x-axis labelled 'v':

```
hist(treecover)
```

---

[6]because of the large amount of data contained in a raster, we may choose to graph a random sub-sample rather than the whole raster using the *maxpixels* argument

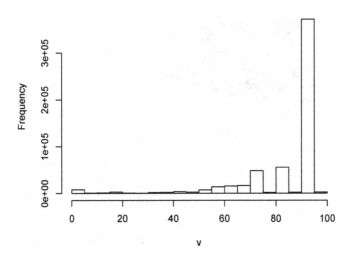

12.12.4 Basic raster calculations

We can get the some basic descriptive data (e.g. sum, mean, min, max, standard deviation) from a raster using the **cellStats()** function in which the *stat* argument is the name of the R function:

```
cellStats(treecover, stat = "mean")

[1] 83.7519
```

We can also use the **summary()** function to get some basic percentile data, including the number of pixels not assigned a value i.e. NAs (which in this example represent the water surrounding the island):

```
summary(treecover)

        waihi_treecover
Min.            0.00000
1st Qu.        84.86244
Median         90.71300
3rd Qu.        93.73967
```

```
Max.            100.00000
NA's       930339.00000
```

12.12.5   Sampling

Often, in conservation projects we want to set up some kind of monitoring based on land use or habitat quality. As a result, we often want to do stratified random sampling[7] based on the value of a variable. We can do this with rasters by first separating the continuous data into groups (called strata) using the **reclassify()** function. In this function we separate the strata based on their minimum and maximum values. Each strata is then assigned a reference number. The general pattern is:

```
reclassify(my.data, c(min.value_1, max.value_1, 1,
                      min.value_2, max.value_2, 2))
```

To make sure we include the lowest and highest values we can use the terms -Inf and Inf. For example, if we were to separate our **treecover** into 3 strata:

```
quality <- reclassify(treecover,
                    c(-Inf, 50, 1,
                      50, 90, 2,
                      90, Inf ,3))
```

We can then use the **sampleStratified()** function to produce a matrix of sample points. For ease of use it is usually easiest to turn the output from a matrix back into a data frame. In the example below, we are choosing 2 random samples from each of the three strata:

```
set.seed(87)
sample <- sampleStratified(quality, size = 2, xy = TRUE)
sample <- as.data.frame(sample)
sample

      cell         x         y layer
1    85211  146.8794  3.747667     1
```

---

[7]see Chapter 14 for background information

```
2 1199512 146.9075 3.575722      1
3 1142266 147.0058 3.584611      2
4 1233307 146.7950 3.570444      2
5  990752 146.9186 3.607944      3
6  803790 146.9847 3.636833      3
```

We can then overlay the selected points onto our plot using the **points()** function from base R by subsetting our data frame for the columns containing the longitude (**x**) and latitude (**y**) coordinates. While we could make a ggplot to do this (see section 12.12.12) base R is substantially faster:

```
plot(quality)
points(sample[ , 2:3])
```

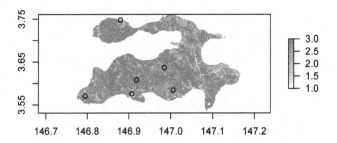

12.12.6   Extracting raster data from points

We can also extract the values of a raster from beneath a data frame containing point data using the **extract()** function. In the example below, we add it to our existing data frame:

```
sample$cover <- extract(treecover, sample[ , 2:3])
sample

    cell       x        y layer      cover
1  85211 146.8794 3.747667     1  0.7865975
```

```
2 1199512 146.9075 3.575722     1 45.1546758
3 1142266 147.0058 3.584611     2 84.8624444
4 1233307 146.7950 3.570444     2 68.8728447
5  990752 146.9186 3.607944     3 93.7396747
6  803790 146.9847 3.636833     3 93.7396747
```

12.12.7 Turning data frames into rasters

Often we will have coordinate data for sample points and we want to turn them into rasters. This can be done via the **rasterize()** function using the dimensions of a pre-existing raster. Only the points used in the function will have raster values – all others will be **NA**. As a result, you generally won't be able to see these points on a raster plot because they will be only a tiny fraction of the raster (unless they are very numerous). But despite this they do exist and are useful for subsequent calculations (especially distance calculations). For example, in order to make our **location** data (5 rows of data) into a raster we would use:

```
loc.raster <- rasterize(location[ , 2:3], treecover)
```

If we now plotted **loc.raster** it would appear to our eyes that the raster is missing – despite this our 5 pixels are there, somewhere.

12.12.8 Calculating distances

The **distance()** function in **raster** is a powerful way to calculate average distances in 2-dimensions. For example, we can use this function to work out the average distance between villages on Waihi using **loc.raster** that we made in the example above:

```
dist.village <- distance(loc.raster)
plot(dist.village)
```

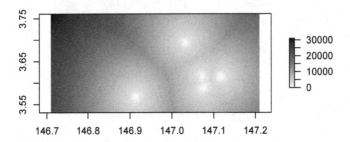

But we see is that the raster that has been created is for the full extent of the raster, not just the island. In order to make this raster appear sensible we must mask it using the another raster.

12.12.9 Masking

In order to match our distance raster to the original outline of the island we can we need to mask our raster by the original **treecover** raster. To do this we use the **mask()** function:

```
dist.waihi <- mask(dist.village, treecover)
plot(dist.waihi)
```

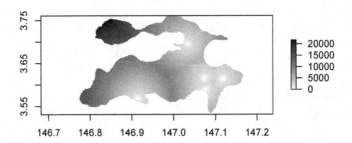

Now we have a good visualisation of straight-line distances between villages. Not only that, but thanks to the other **raster** functions we have learnt how we can now estimate and extract straight-line distances from villages, health clinics, roads, water sources, forest edges, and park boundaries.

12.12.10 Cropping

Sometimes we will want to crop the raster to focus on a smaller area. We can do this using the **crop()** function provided we have established the extent (rectangular boundary) we are interested in:

```
new.extent <- extent(c(xmin = 147.0, xmax = 147.2,
                       ymin = 3.531, ymax = 3.7))

closeup <- crop(treecover, new.extent)

plot(closeup)
```

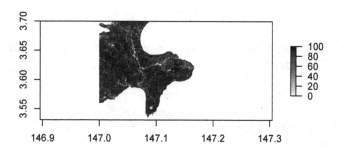

12.12.11 Saving

Rasters can be saved using the **writeRaster()** function. A variety of formats can be used (including GeoTIFF through the argument *format = "GTiff"*). However, the default format for saving rasters in R is as a .grd file. As producing rasters is often time consuming saving a raster you have worked on is sensible. As a precaution against mistakenly overwriting a raster you will have to use the *overwrite = TRUE* argument if you want overwrite a previously saved raster file:

```
writeRaster(dist.waihi,
            filename = "dist.waihi.grd",
            overwrite = TRUE)
```

12.12.12   Changing to a data frame

Rasters can be changed to data frames using the **as.data.frame()** function
so long as the argument *xy = TRUE* is included:

```
new.df <- as.data.frame(dist.waihi,  xy = TRUE)
```

Once changed to a data frame the data can be graphed as a ggplot using
the **geom_raster()** function. **Warning:** because of the large amount of data
contained in many rasters this can take a long time to run:

```
ggplot()+
  geom_raster(data = new.df, aes(x = x, y = y, fill = layer))+
  scale_fill_gradient(low = "grey90",
                      high = "black",
                      na.value = "white",
                      name = "distance (m)")+
  labs(title = "Straight line distance from villages")+
  coord_equal()+
  theme_void()
```

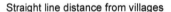

Straight line distance from villages

12.13　Recommended resources

- Robin Lovelace, Jakub Nowosad, and Jannes Muenchow (2019) Geocomputation with R. CRC Press, New York. A free online version is available at: `https://geocompr.robinlovelace.net/`.

- Edzer Pebesma and Roger Bivand (2019) Spatial Data Science. The book currently only exists as an online version available at `https://www.r-spatial.org/book/`

- `https://r-spatial.github.io/sf/`. The official website for the `sf` package.

- `https://rspatial.org/raster/`. The official website for the `raster` package.

12.14　Summary

- One of the most important pieces of information required in any project report is a map.

- Spatial data is data which represents a location in space. There are two main types of spatial data: vectors and rasters.

- Vector data (points, lines, and polygons) are easy to map using the `sf` package as a ggplot.

- Raster data is best handled through the `raster` package.

13

Common R code mistakes and quirks

In this chapter we will look over some of the most common coding mistakes to make, and some R quirks:

- Capitalisation mistakes.

- Forgetting brackets, quotation marks, and commas.

- Forgetting '+' in a ggplot.

- Forgetting to call a ggplot object.

- Piping but not making an object.

- Changing a factor to a number.

- Variables automatically being read as factors.

13.1 Making mistakes

Everybody makes mistakes and this is part of the learning process. When you are learning R you will make more mistakes than when you are more experienced. But even when you are experienced you won't ever stop making mistakes – you'll just get better at spotting and correcting them.

Our brains are very good at reading and filling in the gaps – usually we can still understand what a piece of text is trying to explain even if there are spelling and punctuation mistakes. Code, by comparison, is a set of instructions. If an instruction is not recognised by the software an error will result. Error messages are there to help us understand why the code cannot be run. Errors in R can be frustrating when we are learning – because the R error messages are not easy to understand. In the following sections, I outline some of the most common problems, what R is expecting, and how you can correct the mistake. I also cover a couple of situations when the code runs but gives us an expected output.

13.2   The packages

If you want to replicate the error messages associated with making a ggplot
you will need to load the `tidyverse` package:

```
library("tidyverse")
```

13.3   The data

Some of the examples in this chapter use the `demo1` data set from the `condev`
package. If you want to replicate the errors you will have run the following
code:

```
library("condev")
data(demo1)
```

13.4   Capitalisation mistakes

```
x <- 7
X
```

```
Error: object 'X' not found
```

R is case sensitive. Small mistakes in capitalisation will mean that R will
not be able to find our objects. This situation can be avoided by always using
lowercase lettering for objects, and making sure object names are distinctive.
It is easy to make capitalisation mistakes – and we often won't notice the
mistake until we receive an error.

13.5 Forgetting brackets

```
mean(c(6,18,19)
```

Failing to close the brackets of a function will leave a plus sign hanging in the console window:

```
+
```

Missing a bracket is very easy to do if we are not using an integrated developer environment (like RStudio) which can pick up these kinds of minor mistakes automatically. In this situation R is simply expecting more information before it can run the code. As a result a '+' remains in the console window.

To get out of this situation write and run the missing bracket. Alternatively, simply push the escape key 'ESC' on the keyboard, close the bracket, and rerun the code. If the code has been executed successfully the prompt sign should reappear:

```
>
```

Now we are all good to continue.

13.6 Forgetting quotation marks

```
ggplot()+
    geom_histogram(data = demo1, aes(x = x, fill = group1),
                    position = dodge)

Error in layer(data = data, mapping = mapping, stat = stat,
geom = GeomBar,   : object 'dodge' not found
```

This message indicates we have forgotten to use quotation marks around the *dodge* argument – as a result the function expects 'dodge' to be an object (which it is not). The use of quotation marks can vary between functions and packages[1]. In most R functions quotation marks are expected when an argument expects you to choose from a set of text-based options. However, you should check the function documentation using the R help (?) function to make sure.

13.7   Forgetting commas

```
ggplot()+
  geom_histogram(data = demo1, aes(x = x fill = group))

Error: unexpected symbol in:
"ggplot()+
  geom_histogram(data = demo1, aes(x = x fill"
```

It is easy to forget a comma (or substitute it with a full stop) in a function. Either way will result in the same error – telling you there is an unexpected symbol. In such cases we just have to go over our code and try and spot the missing comma.

13.8   Forgetting '+' in a ggplot

```
ggplot()
  geom_point(data = demo1, aes(x = x, y = y))+
  theme_bw()

Error: Cannot add ggproto objects together.
Did you forget to add this object to a ggplot object?
```

---

[1]for example **library()** and **data()** functions can both work with or without quotation marks

This message indicates we have forgotten a '+' sign between ggplot layers (in this example at the end of the first line). We need to go back and search our ggplot code for a missing '+' sign, replace it, and rerun.

13.9   Forgetting to call a ggplot object

```
my.graph <- ggplot()+
    geom_point(data = demo1, aes(x = x, y = y))+
    theme_bw()
```

This is one of my all time favourite mistakes. You are waiting for a ggplot to appear and the graph never appears. This happens when we get in the habit of making ggplot graphs without making them into objects. Absolutely nothing is wrong with the code. All we have done is forgotten to call the object in order to see the graph. The key line we needed to run was:

```
my.graph
```

13.10   Piping but not making an object

Let's imagine we have been using a pipe (%>%) to run a series of functions:

```
demo1 %>%
    group_by(group1) %>%
    summarise(average.y = mean(y))

# A tibble: 3 x 2
  group1 average.y
  <fct>      <dbl>
1 A           8.27
2 B           9.15
3 C           9.97
```

We can see the output in the console. Everything looks great. But when we call **demo1** nothing appears to have changed (first 6 rows of data shown):

```
demo1

            x          y group1 group2         error
1 10.142924 10.251406      C treat1 -0.14015255
2  9.962386  9.626123      C treat1  0.34827946
3 10.328865 10.395164      C treat1 -0.35885953
4  9.566731  9.531268      C treat2  0.01146037
5 10.344349  9.873001      C treat2 -0.12696158
6  9.859476  9.348461      C treat2  0.68017090
```

The key here is we are not overwriting the object **demo1**. If we want to do that we need to use the assignment operator '<-':

```
demo1 <- demo1 %>%
    group_by(group1) %>%
    summarise(average.y = mean(y))
```

13.11   Changing a factor to a number

Let's create a variable (**alive**), which looks like a number but is actually is a factor. A classic example is using '0' and '1' to represent two possible states[2]:

```
alive <- as.factor(c(0,1,0,1,0,1,1))
class(alive)

[1] "factor"
```

Using base R our natural inclination would be to just force it to be numeric

---

[2]be careful using '0' and '1' as factors in data frames. As these names are not informative there is a risk you might not notice a conversion mistake and then misinterpret the output of your analysis

using the **as.numeric()** function. But on inspection it doesn't do what we want:

```
as.numeric(alive)

[1] 1 2 1 2 1 2 2
```

This is one of the famous quirks of R: the **as.numeric()** function turns the factors into their factor levels (i.e. 1 & 2) rather than their values. To get around this we must first get R to recognise the values as characters using the **as.character()** function before changing them to being numeric:

```
alive <- as.numeric(as.character(alive))
alive

[1] 0 1 0 1 0 1 1
```

13.12   Strings automatically read as factors

Prior to R version 4.0, if we used the **read.csv()** function any column with any text or a symbol in it was automatically turned into a factor by R. This made perfect sense when when R was primarily used by statisticians and the data was expected to be entered without mistakes. However, these days R is used a lot for data wrangling and cleaning. As a result, we often need to change the text and consequently, R was changed in version 4.0 to import columns of text without changing them into factors. However, the issue with factors will persist for people using earlier versions of R. This is a problem because factors are harder to change because extra information is coded into pre-defined levels. We can demonstrate this problem if we imagine we have read a .csv file which contains just 3 rows of data[3]:

```
example <- read.csv("example.csv")
```

---

[3]as no such file exists on your computer this example will not run

```
example
```

```
  Ref  Group
1   1 Novera
2   2 Kuange
3   3 Kuange
```

Looking at the data we might suddenly realise that the second observation in `Group` shouldn't be 'Kuange' but 'Arihafa'. So we try to change it using base R:

```
example$Group[2] <- "Arihafa"
```

```
Warning message:
In `[<-.factor`(`*tmp*`, 2, value = c(2L, NA, 1L)) :
  invalid factor level, NA generated
```

We get this error message because 'Arihafa' isn't one of the two original factor levels within the `Group` variable (i.e. 'Novera' or 'Kuange'). We can avoid this situation when importing data with older versions of R by:

1. using the argument *stringsAsFactors = FALSE* with the **read.csv()** function:

   ```
   example <- read.csv("example.csv",
                       stringsAsFactors = FALSE)
   ```

   or

2. using the **read_csv()** function from the `tidyverse`:

   ```
   library("tidyverse")
   example <- read_csv("example.csv")
   ```

Both methods will result in these columns containing text or symbols being read as characters. In this way the issues associated with modifying factors can be avoided.

13.13 Summary

- Most R code errors are a result of capitalisation and punctuation mistakes.

- The simplest mistakes we make are forgetting to call an object or not making a new object in the first place.

- The most troublesome R quirks tend to involve factors. However, automatic conversion of strings into factors is no longer the default for R versions $\geq$ 4.0.

Part III

Modelling

14

Basic statistical concepts

In this chapter we will introduce some basic statistical terms and concepts including:

- How to describe things which are variable.

- Probability and probability distributions.

- An overview of modelling approaches.

14.1 Variables and statistics

As people, we spend a good deal of our lives naming and describing things. But how do we describe a thing which is variable in value – like how tall a group of people is? Fortunately, mathematics has given us a given special set of terms which can help us. The first term we need to understand is 'variable' – which refers to anything that is variable in value (e.g. in our example, height). In this chapter we will explore other terms which help us describe patterns seen in variables. Many of these patterns can be described using summarised values which are known as statistics. As a result, we call the mathematical field which studies these patterns – statistics.

14.2 The packages

In this chapter we will need to load the `tidyverse` package:

```
library("tidyverse")
```

14.3 The data

In this chapter we will simulate our data so no pre-loaded data sets are required.

14.4   Describing things which are variable

When we are trying to describe the basic pattern of a variable we need to think about two characteristics:

- a midpoint (e.g. the height of a typical individual in the group).

- the variability around that midpoint (e.g. the variation in the height of the people in group)

The first concept (a midpoint) is known in statistics as central tendency while the second concept (regarding variability) is called variability. As a result, when we describe a variable to other people it is helpful to report measures of both central tendency and variability. Let's examine these concepts by creating an object called `height` (measured in centimeters for 5 people) – we will refer back to this example throughout this chapter:

```
height <- c(167.2, 168.3, 169.1, 170.2, 210.7)
height

[1] 167.2 168.3 169.1 170.2 210.7
```

14.4.1   Central tendency

The two most common methods of reporting central tendency are the mean (also known as the average) and the median.

14.4.1.1   Mean

The mean is the most common measure of central tendency. The mean represents the mid-point of the observations in terms of their value. It is calculated as the sum of the variable's values divided by the number of observations. We can calculate the mean using the **sum()** function and dividing by the **length()** function (which counts the number of observations in a vector):

```
sum(height) / length(height)

[1] 177.1
```

However, a much faster way to get the mean is to use the **mean()** function:

```
mean(height)
```

```
[1] 177.1
```

The mean is a good measure of the mid-point if the observations are distributed symmetrically about the mean. However, sometimes the mean might not be a very representative midpoint for the data. In our example, we can see that the mean is 177.1 which is higher than every observation except the largest:

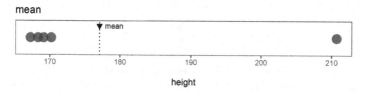

In such cases we might want to consider using a different measure of central tendency such as the median.

14.4.1.2   Median

The median is the midpoint of the observations when ordered from lowest to highest value. If we were to order the 5 observations of **height**, the median would be the third (i.e. 169.1):

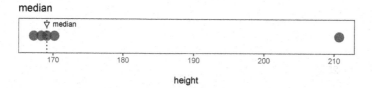

If the number of observations is odd the median will always be a value present in the data set (the middle ranked observation). However, if the number of observations in the data is even, the median is made by averaging the value of the two observations nearest the midpoint (in terms of order). The median of the **height** can be calculated using the **median()** function:

```
median(height)
```

```
[1] 169.1
```

If the mean and median are not similar this suggests that the data is skewed (not symmetrical about the mean). Often data is skewed because it is being influenced by extreme values. In these situations the median is usually a better representation of central tendency.

In our `height` example, below we can see that the mean (177.1) and median (169.1) are dissimilar. This is because the mean is being heavily influenced by a single observation (210.7) which appears to be unusually large. As a result, we say that this data has a large skew:

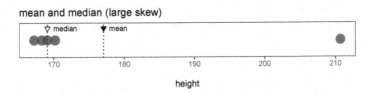

However, at small sample sizes there can be differences between the mean and the median, even if the data is symmetrical (not skewed). In this situation the differences between the mean and median disappear when the sample size (usually referred to in statistics by the letter n) is increased. As a result, in non-skewed data the mean and median usually resemble each other when the sample size is large:

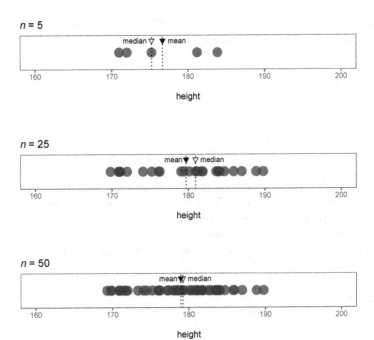

Sample size
When our sample size is small, statistical methods can struggle to detect the underlying pattern in the data. As a result, the first step of designing a study is to make sure our sample size will be big enough to accurately describe the patterns we might expect to see. Additionally, because the precision of statistical methods (how well the technique returns a similar result from different samples) depends upon the sample size we should always report our sample size.

14.4.2   Describing variability

The three most common methods for describing variability in data are the range, standard deviation, and percentile range.

14.4.2.1   Range

The simplest way to describe the variability in data is to use the **range**. The range is simply the span of values between the minimum and the maximum. The issue with this approach is that we are trying describing the data by its most extreme values – which might not be a good representation of the data. The range of the **height** can be calculated using the **range()** function:

```
range(height)

[1] 167.2 210.7
```

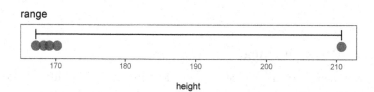

14.4.2.2   Standard deviation

The standard deviation is the equivalent of the mean but for variability[1]. It represents a measure of the distance between each observation of a variable and the mean. Like the mean, the standard deviation can be skewed by extreme values. Consequently, it is most useful for data which isn't skewed. The standard deviation of the **height** can be calculated using the **sd()** function:

---

[1]see Appendix for equation

```
sd(height)
```

```
[1] 18.81502
```

We can see in the non-skewed sample below that the standard deviation appears to be a useful description of the average variability in the data:

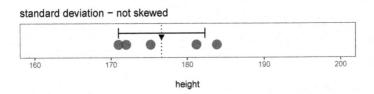

By comparison, the standard deviation doesn't describe the variability at all well when the data is skewed (e.g. our `height` example) as it can exceed the lowest values and underestimate the extent of the variability:

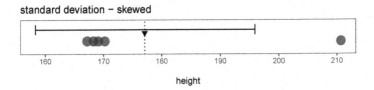

14.4.2.3 Percentile range

Alternatively, we can use percentiles to describe the variability of the data using the **quantile()** function. Percentiles give the expected value of observations for a given proportion (between 0 and 1) of the ordered values e.g. the 2.5th and 97.5th percentile of the height sample would be:

```
quantile(height, c(0.025, 0.975))
```

```
  2.5%  97.5%
167.31 206.65
```

Because our sample is small, it doesn't include enough observations for R to find these percentiles from the sample order. As a result these values have to be estimated by the function. When we look at the 5th–95th percentile range of our skewed `height` data we can see it avoids the nonsensical values of the standard deviation:

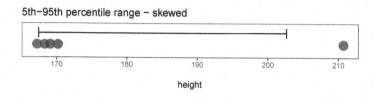

It also works well for non-skewed data:

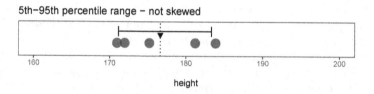

As percentile ranges work quite well on both skewed and non-skewed data, they are a useful tool for exploring variability in our data. Indeed, percentiles form the basis of box plots (see Chapter 7) in which the box portion of the plots are constructed from the median and the 25th and 75th percentiles. Large asymmetries in the length of a box may alert us to issues of skewness. For this reason, box plots should be one the first tools to we use when visualising data.

14.4.3 Reporting central tendency and variability

Generally, we want to present both the measure of central tendency and variability when describing a variable. Because there are a number of different methods of reporting variability we need these to be obvious to the reader. Different reporting and abbreviation styles exist (with the ± sign recently falling out of favour — see below). If our results are being published we should follow the preferences of the publisher overseeing the report. Below are examples of some ways of presenting these descriptive statistics:

Reporting mean with standard deviations

- *Contemporary*: Mean height was 177.1 cm (sd = 18.8 cm)

- *Traditional*: Mean height was 177.1 ± 18.8 cm (mean ± SD)

- *Bracketed in a sentence*: (mean = 177.1 cm, sd = 18.8 cm)

Reporting median with range or percentile range

- Median height was 169.1 cm (range = 167.2 – 210.7 cm)

- Median height was 169.1 cm (5th–95th percentile range = 167.4 – 202.6 cm)

14.4.4   Precision

A general rule of statistical reporting is that our final output should not be more precise than our raw data. The precision of our data is determined by the number of digits it contains. In most circumstances we should round our final results to match the number of digits in our original data. If we look at our `height` data set we can see we reported the raw data with 4 digits yet the **quantile()** function gave an answer with 5 digits, and the **sd()** function gave an answer with 7 digits. As a result, if we were reporting these results we should round them to 4 digits to match the original data. Some people (and fields) advocate additional rules for the number of digits in statistical outputs. If you are following such rules keep in mind any rounding should only be done to the final output – not your intermediate workings.

14.5   Introducing probability

If you remember back in Chapter 2 we introduced the concept of inductive reasoning. This is the idea that a conclusion can never be certain, it can only be probable. When we talk about probability we are talking about the chance an event could happen. We can write probability as a proportion using the following equation:

$$\text{Probability of an event happening} = \frac{\text{Number of ways it can happen}}{\text{Total number of possible outcomes}}$$

For example, in a coin toss – there are only two sides (a head, and a tail) – so there are only two possible outcomes. But only one side of the coin has a head. Therefore, the probability of getting a head from one toss is:

$$\text{Probability of a getting a head} = \frac{\text{Number of sides with a head}}{\text{Total sides of a coin}}$$

$$\text{Probability of a getting a head} \quad = \frac{1}{2} = 0.5$$

However, the actual outcome of a coin toss, is subject to chance. So, if we were to make one thousand coin tosses we shouldn't expect to get exactly 500 heads and tails. However, if we were to do one thousand tosses and got only 119 heads – we might suspect that the coin toss was not fair. Fortunately, statisticians have developed tests to help us understand how likely such outcomes are. Most, but not all, of these statistical tests are based on the understanding of probability distributions[2]. As a consequence, it is important to have a basic understanding of the main probability distributions. Not only will this knowledge guide us to use the appropriate statistical test, but it will allow us to simulate the right kind of data sets at the Plan phase of the Deming cycle (see Chapter 3).

14.6   Probability distributions

Probability distributions guide how we approach statistical modelling. A probability distribution is a mathematical function that gives the expected probabilities for different possible outcomes. The most common distributions we come across in conservation and development projects are:

- The Bernoulli distribution (a type of binomial distribution).

- The Poisson distribution.

- The normal distribution.

14.6.1   Binomial distribution

The binomial distribution is important for modelling a variable that can only result in two possible outcomes (i.e. binary outcomes). Below are some examples:

- 'present' or 'absent'

- 'alive' or 'dead'

- 'yes' or 'no'

- 'heads' or 'tails'

---

[2]tests which are not based on probability distributions are called non-parametric tests

Let's return to our coin toss. If we were to toss a coin once, our expected outcome is a head or a tail – each with the probability of 0.5. If we were to toss the coin 8 times (8 trials) then, on average, we would expect 4 of the outcomes to be heads. Similarly, if we were to toss the coin 30 times (30 trials) on average, we would expect 15 of the outcomes to be heads. However, random chance means that there will be variability around that number. We can visualise this variability in Figure 14.1. We can see that the distribution changes as the number of trials increases – becoming less rectangular and more bell-shaped as the sample size increases.

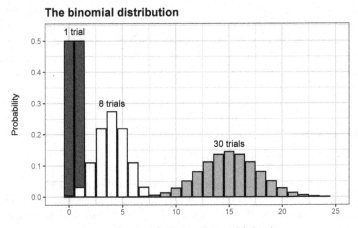

FIGURE 14.1
The binomial distribution

14.6.1.1   Bernoulli distribution

In conservation and development projects the most common type of binomial modelling involves a situation where the number of trials = 1 (representing a one time measurement of success or failure). This is known as the Bernoulli[3] distribution which is a special case of the binomial distribution. For example, if we wanted to simulate a Bernoulli distributed variable (e.g. the number of turtle eggs surviving into juveniles) we could do this using the **rbinom()** function. This function has three arguments:

- n: the number of observations (the number of eggs)

- *size*: the number of trials (1)

- *prob*: the probability of success (the probability of surviving i.e. being found 'alive')

---

[3]named after Jacob Bernoulli – a 17th century mathematician

A good practice in R simulations is to use the **set.seed()** function to ensure others (including our future selves can replicate the results). Now, lets use the **rbinom()** function to simulate the number of turtle eggs surviving into juveniles from two different nests (with different starting numbers of eggs, and different probabilities of survival):

```
set.seed(13)
survival.nest1 <- rbinom(n = 83, size = 1, prob = 0.30)
survival.nest2 <- rbinom(n = 70, size = 1, prob = 0.40)
```

Each line of the **rbinom()** code above results in a vector in which 0 = failure (didn't survive), and 1 = success (survived) e.g. for `survival.nest1`:

```
survival.nest1

[1]  1 0 0 0 1 0 0 1 1 0 0 1 1 0 0 0 0 0 1 0 0 0 0 0 0 0 0 0
[29] 0 0 0 0 1 1 0 0 0 0 1 0 0 0 1 0 0 1 0 0 1 0 0 1 0 0 0 0 1 0 0
[57] 0 0 0 0 0 0 1 0 0 0 0 0 0 0 0 0 0 0 0 1 0 0 0 0 0 0 1
```

As a Bernoulli distribution has only two possible outcomes the best way to graph these results are as a stacked bar graph. We can do this with a ggplot using the **geom_bar()** function. But first we will have to make a small data frame:

```
survival <- c(survival.nest1, survival.nest2)
nest <- c(rep("turtle 1", 83), rep("turtle 2", 70))
bern.sim <- data.frame(survival, nest)
```

We can quickly summarise our `survival` data set using the **table()** function. This shows us the number of eggs surviving to become juveniles in each turtle's nest:

```
table(bern.sim$survival, bern.sim$nest)

    turtle 1 turtle 2
0        66       40
1        17       30
```

In order to get the stacked effect in a bar graph we will have to ensure we use the *fill* argument within the **aes()** function. As the *fill* argument only works with factors we have to change `survival` using the **as.factor()** function either within the ggplot (or by modifying the vector in the original data frame). For visualising binary data it is also helpful to show proportionality by using the *position = "fill"* argument:

```
ggplot()+
  geom_bar(data = bern.sim, aes(x = nest,
                                fill = as.factor(survival)),
          position = "fill") +
  labs(y = "Proportion of total eggs",
       x = "Site")
```

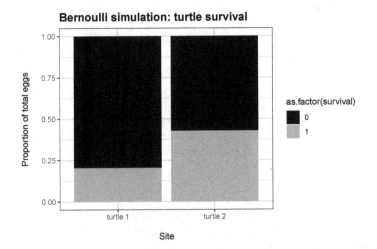

This graph can help us visualise how many turtles died (0) compared to those who survived (1). Stacked bar graphs like this are especially useful when we are trying to compare Bernoulli outcomes across multiple groups.

14.6.2   Poisson distribution

A different type of distribution is used when dealing with count data. Count data can only be made up of whole numbers (e.g. 0, 1, 2, 3...). Consider the following types of count data:

- The number of patients visiting an emergency clinic each day.

- The number of animals appearing at a water hole.

- The number of soldiers being killed each year by being kicked in the head by a horse[4]

All of these situations are examples of the Poisson distribution. The Poisson distribution is used for count data and models the number of times an event occurs in interval of time (e.g. number of visits in a day) or space (e.g. number of individuals in an area). The Poisson distribution has some key assumptions:

- – events occur independently (e.g. patients visiting a clinic don't attract or repel other patients).

- – events do not do not occur at exactly the same time (although they can be very close together).

The Poisson distribution changes shape as the mean number of events increases moving from a curve with a 'shoulder' to a more bell-shaped distribution (Figure 14.2). In the Poisson distribution the mean and its variability (as measured by variance, the standard deviation squared) are the same – and are usually referred to as lambda[5]. We can simulate random Poisson data using the **rpois()** function. This function has 2 arguments:

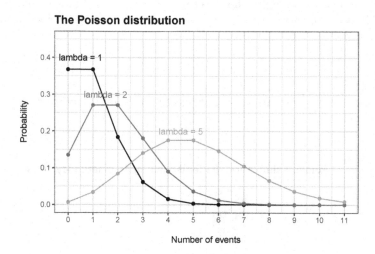

FIGURE 14.2

The Poisson distribution

- *n*: the number of observations.

- *lambda*: the mean count.

---

[4]the most famous example of the Poisson distribution described by Ladislaus Bortkiewicz

[5]as a result, a critical check of count models based on the Poisson distribution is that the mean equals the variance

For example, if we can simulate the number of patients coming to a village clinic over a 90-day period based on a mean of 5 visits per day:

```
set.seed(13)
patients <- rpois(n = 90, lambda = 5)
patients
```

Which results in the following vector:

```
 [1]  6  3  4  2  9  1  5  7  8  2  6  8  8  5  5  4  4  5
[19]  7  6  3  5  6  5  2  6  1  5  4  6  4  4  8  7  5  2
[37]  5  2  6  2  5  3  9  5  6  8  3  4  6  6  5  4  2 12
[55]  1  5  3  4  3  2  5  5  7  6  4  5  4  2  4  4  4  4
[73]  4  3  4  9  3  5  3  6  5  5  9  4  6  5  9  4  4 14
```

We can then graph the vector's data as a ggplot using the **geom_histogram()** function (note: by using the *colour = "white"* argument we can make the edges of the bars clearer):

```
ggplot() +
    geom_histogram(aes(x = patients), colour = "white",
                   binwidth = 1)
```

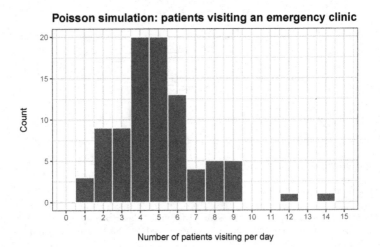

The importance of recording zeros

A common problem encountered in conservation and development data is that people often forget to record observations when the count is zero. This creates a number of problems. For example, if we are given someone's notes regarding the number of elephants seen near a village everyday, if the person only recorded elephants when they were present what does it mean if no elephants were recorded? Does it mean zero elephants were seen, or does it mean that the person did not make a record that day? While we might be able to ask the record keeper directly, sometimes this is difficult or even impossible. Additionally, if people fail to record zeros, often those analysing the data will not realise and just use the counts they are given – the result being they will overestimate the count, and as a consequence their analysis will be incorrect.

14.6.3   Normal distribution

The normal distribution (sometimes called a Gaussian distribution) is always bell-shaped and symmetrical. It is the most widely used distribution in statistical modeling. Interestingly, if we repeatedly sample a truly random variable we will find that the distribution of its sample means (and sums) will approximate the normal distribution. Therefore, variables which are the sum of many independent processes can be expected to have distributions that are, more-or-less, normal distributed. Consequently, the normal distribution can be expected to fit many natural phenomena. As a result, many statistical tests are based around the assumption that the data are normally distributed. The exact shape of the normal distribution is defined by the mean and standard deviation of the variable we are measuring. Furthermore, because of the distribution's symmetry we can expect:

- $\pm$ 1 standard deviation will hold ~68% of observations

- $\pm$ 2 standard deviations will hold ~95% of observations

- $\pm$ 3 standard deviations will hold ~99.7% of observations

We can see this visually in Figure 14.3.

Consequently, any observation further than $\pm$ 3 standard deviations from the mean is highly unusual. The exception is when the sample size is very large – in which case we could expect approximately 0.3% to exceed 3 standard deviations. When extreme observations are unexpected we call them 'outliers'. As a result, it is normal practice to investigate them (especially to ensure they are not mistakes).

We can simulate normally distributed random data using the **rnorm()** function. In the function we need to give arguments for: number of samples (n), their mean (*mean*), and their standard deviation (*sd*):

1 standard deviation

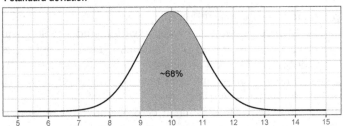

2 standard deviations

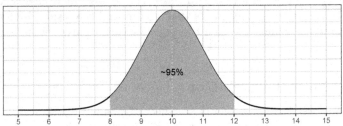

3 standard deviations

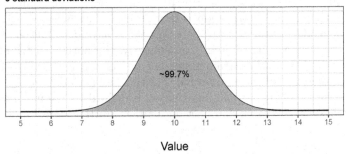

Value

FIGURE 14.3
Observations contained within the bounds of different standard deviations
under the normal distribution for a variable (mean = 10, sd =1)

```
set.seed(13)
rnorm(n = 10, mean = 10, sd = 1)

[1] 10.554327  9.719728 11.775163 10.187320 11.142526
[6] 10.415526 11.229507 10.236680  9.634617 11.105144
```

When the sample size is small the shape of the distribution is unclear.

Only when the sample size is greatly increased can we see the bell-shape of the normal distribution develop (Figure 14.4).

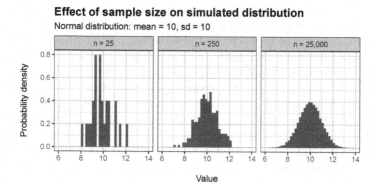

FIGURE 14.4
The effect of sample size on the normal distribution

14.7   Random sampling

In statistics we use samples to make inferences about a population. The sample is what we measure, but the population is what we are trying to describe. For example, we may capture and weigh 50 wild tigers (the sample) to better understand the size of the wild tigers alive today (the population). Usually the size of the population is so great that we cannot possibly measure all of its members[6]. So normally, we have to sample – which means examining just a subset of the whole population. Almost all statistical methods are built around the assumption that such samples are drawn randomly. By sampling randomly we can avoid bias (i.e. getting samples which are not representative of the population). Some examples of bias in conservation and development could be:

1. Trying to understanding the views of people in a village by only surveying men aged between 40 and 50 years old.

---

[6]however, there are some endangered species which have so few living individuals it is possible to measure the entire population

2. Only surveying people who have proactively volunteered to give their opinions.

3. Estimating the density of forest elephants by only taking measurements near road sides.

Now, let's consider why these samples might be biased. In the first example, middle aged men will be over represented while every other villager is excluded. In the second example, by having people self-select we are restricting our sample to people with outgoing personalities and excluding shy and introverted people. In the third example, the resulting elephant density risks being only representative of road side environments – not the greater forest.

Avoiding bias does not come naturally to people. People tend to choose samples which are easy to obtain[7] which will usually result in having a biased sample. The key to avoiding sampling bias is choosing the sample subjects and locations randomly, and in advance. The two most common types of random sampling used conservation and development are: simple random sampling and stratified random sampling.

14.7.1   Simple random sampling

In simple random sampling we just want to choose samples randomly. This type of sampling is straight forward. To demonstrate, let's make a data frame (`our.data`) consisting of 1,000 potential samples. We will use the **paste()** function to make 1,000 'names' – these could represent people (or households) in a village who we might want to interview:

```
our.data <- data.frame(potential.samples =
                    paste("P_", 1:1000, sep = ""))
```

As our sampling is random (but we may need to reproduce it later) we can use the **set.seed()** function to make the results reproducible. This way, anyone who uses the same seed will get the same results. Our next step is to make an index of the rows of our data. To do this, we first count the number of rows using the **nrow()** function and make an object (`count.rows`). We then use the **sample()** function to select a random number of rows using the *size* argument, and make this an object (`index`). Importantly, we must use the *replace = FALSE* argument to ensure samples do not get repeated. We then subset `our.data` by `index` to get a random sample of rows from the data frame:

---

[7]this is commonly known as convenience sampling – which is not a type of sampling but a criticism of the method used

```
set.seed(17)
count.rows <- nrow(our.data)
index <- sample(count.rows, size = 10, replace = FALSE)
our.sample <- our.data[index, ]
our.sample

[1] P_946 P_993 P_734 P_108 P_495 P_247 P_941 P_423 P_810
[10] P_150
1000 Levels: P_1 P_10 P_100 P_1000 P_101 P_102 ... P_999
```

14.7.2   Stratified random sampling

A common alternative to simple random sampling is to use stratified random sampling. Stratified random sampling is useful when we know that certain groups (called strata) already exist within our sample and we would like to randomly sample from each of these strata to ensure our sample is representative.

Consider the following example: we are planning to set up camera-traps to count tigers in a protected area and we need to choose sampling locations from an established list. However, we know the area is made up of three very different habitats and these affect tiger density. In this situation our strata would be the different kinds of habitat. We can simulate this by adding a new variable (called **strata**) to **our.data** using the **sample()** function. In this example, we have three types of habitat ('forest', 'woodland', and 'grassland'), however, the 'grassland' habitat is twice as common as the others:

```
our.data$strata <- sample(c("forest", "woodland",
                            "grassland", "grassland"),
                          size = 1000, replace = TRUE)

table(our.data$strata)

  forest grassland  woodland
     277       473       250
```

In this example, we can imagine that **our.data$potential.samples** now represents potential locations for camera-traps. In order to do stratified random sampling we have to randomly sample from each strata. We can use **tidyverse** functions to do this for us – using the **group_by()** function in combination with the **sample_n()** function to get a random sample of the same size from each strata (restricted to 3 for display purposes):

```
our.data %>%
  group_by(strata) %>%
  sample_n(3)

# A tibble: 9 x 2
# Groups:    strata [3]
  potential.samples strata
  <fct>             <chr>
1 P_797             forest
2 P_939             forest
3 P_46              forest
4 P_160             grassland
5 P_187             grassland
6 P_736             grassland
7 P_162             woodland
8 P_482             woodland
9 P_275             woodland
```

Alternatively, we could use the **sample_frac()** function to get a sample which is proportional to the size of each strata (in the example we want our total sample to be 1% of the total observations):

```
our.data %>%
  group_by(strata) %>%
  sample_frac(0.01)

# A tibble: 10 x 2
# Groups:    strata [3]
  potential.samples strata
  <fct>             <chr>
1 P_429             forest
2 P_572             forest
3 P_243             forest
4 P_767             grassland
5 P_282             grassland
6 P_512             grassland
7 P_968             grassland
8 P_605             grassland
```

```
 9 P_896            woodland
10 P_230            woodland
```

14.8   Modelling approaches

Often we want to move beyond simply describing the central tendency and variability of our data and develop an understanding of how the world works. The missing link between the two is an ability to make accurate and consistent predictions from the data we have gathered. In particular, in conservation and development we want to use this ability to build more effective interventions.

At the simplest level, we are interested in the prediction of some key variable which is of importance to our project. We call the variable we want to predict as the response variable (also known the dependent variable). Mathematically what we are wanting to do is predict the response variable using an explanation involving one or more other variables which we call explanatory variables (often called independent variables).

Most people are very good at thinking of a variety of different explanations. We call such explanations hypotheses, and the process of making them hypothesising. As we shall see in the next chapter hypotheses can be written as mathematical formulas – which we usually call models. While hypothesising is easy, finding the best model which explains the response variable is a harder challenge. Complicating the situation is that there are a number of different approaches to finding the best model. In the following sections, I briefly describe some of the modelling approaches which can be used to address this challenge.

14.8.1   Null hypothesis testing

Null hypothesis testing was the dominant modelling approach in the 20th century. Null hypothesis testing focuses on assessing whether or not a model is a good explanation based on the chances of getting the same outcome (or one equally rare or rarer) by chance. As a consequence, null hypothesis testing only considers two models:

1. a 'null hypothesis' in which the model lacks a particular explanatory instruction.

2. an 'alternative hypothesis' which includes the extra explanatory instruction.

In null hypothesis testing the null hypothesis is either rejected (suggesting support for the alternative hypothesis) or fails to be rejected (meaning there is little support for the alternative hypothesis). Null hypotheses are always stated in the negative e.g.:

- 'There is no difference between the proportion of men and women attending our training courses' (i.e. gender is not in the model).
- 'The weight of the harvested fish is not associated with the type of fish food they were fed' (i.e. type of fish food is not in the model).
- 'There is no relationship between the month and the number of the pest species caught in traps' (i.e. month is not in the model).

The assessment of the null hypothesis is based on the overall p value produced by the test on the alternative hypothesis. A p value measures the probability of getting an outcome equally rare, or rarer, by chance, if the null hypothesis was correct. Traditionally, if a test resulted in a p value equal to or less than a probability of 0.05 (a 1 in 20 chance) the null hypothesis was rejected and the result was termed 'significant' (meaning the alternative hypothesis was considered a better explanation). On the other hand, if a test resulted in a p value larger than 0.05 then the null hypothesis would not be rejected (and the test would be termed 'not significant').

Currently, the use of null hypothesis testing is falling out of favour. Unlike other modelling approaches null hypothesis testing doesn't allow for multiple alternative hypotheses to be compared against each other simultaneously. Critics also point out that the use of a hard threshold for determining significance (i.e. p value ≤ 0.05) is entirely arbitrary, it promotes focusing on the concept of significance and not the implications of the model, and that the general approach is not intuitive. Additionally, as significance is a function of sample size a significant result can usually be obtained by increasing the sample size. Therefore, a significant result itself doesn't guarantee the result is meaningful e.g. a significant test result could show that the height of two groups of people might be different but the actual difference between the groups could be so small as to be meaningless.

However, as null hypothesis testing dominated statistical testing in the 20th century you will frequently come across the approach. Additionally, many statistical tests in R rely on p values. Consequently, you will need some understanding of them.

What is the difference between probability and a p value?
Let's imagine tossing a coin three times:

- **probability**: what is the probability of getting 3 heads? Given that there are two outcomes (a head or a tail) but only 1 of them can be a head, the probability of getting a head on a single coin toss is $1/2 = 0.5$. Therefore the probability of getting 3 heads would be $0.5 \times 0.5 \times 0.5 = 0.125$.

- **p value**: what is the probability of an outcome as rare, or rarer than getting 3 heads? As shown above the probability of getting 3 heads is 0.125, but the probability of getting 3 tails is also 0.125. As both outcomes as as rare as each other (and there is nothing rarer) the p value is 0.125 + 0.125 = 0.25.

As mentioned previously, one of the biggest problems with null hypothesis testing is that it doesn't allow multiple hypotheses to be compared simultaneously. By comparison, the three approaches outlined in the next sections: information-theoretics, Bayesian inference, and machine learning all allow multiple models to be tested.

14.8.2   Information-theoretics

In information-theoretics we develop a set of candidate models (hypotheses) to be compared. These candidate models are then ranked in terms of the quality of their explanation in terms of Akaike's Information Criterion (AIC)[8]. The quality of the explanation is based on the goodness-of-fit of each model and the number of parameters (see the explanatory box at the end of this section).

AIC is calculated through an equation that measures goodness-of-fit and penalises the number of estimated parameters (the statistical instructions). In information-theoretics the candidate models are ranked according to their AIC (lower being better) — in this way the best model of the group can be identified[9]. The process of ranking models is known as model selection. As information-theoretic methods are very easy to implement Chapters 15 and 16 focus on this approach.

Goodness-of-fit and parameters: an analogy
Consider the following situation. Friend A gives you instructions for getting to the market. They give you a set of two instructions. You follow their instructions and find the market. The next day you have forgotten the way and ask Friend B. They give you a set of twenty instructions. You follow their instructions and find the market. Which explanation is better? Clearly, that of Friend A because it was sufficient, yet shorter.

Consider an another situation. Friend A gives you instructions for getting to the market. They give you a set of two instructions. You follow their instructions and find the market. The next day you have forgotten the way and ask Friend B. They also give you a set of two instructions. You follow their

---

[8]named after Hirotugu Akaike who published the information criterion in 1973

[9]this is often equated to Occam's razor and the principle of parsimony. Both concepts are broadly based on the idea that an explanation should be no more complex than it needs to be

instructions and get lost only 100m before the market. Which explanation is better? Clearly, that of Friend A because despite both explanations having the same number of instructions only that of Friend A was good enough to get you to the market. If you think about it, even if Friend A gave you forty instructions it would still be better than the explanation of Friend B – because despite being longer they were the only instructions that got you to the market.

These examples roughly match the statistical concepts of parameters and goodness-of-fit. Parameters can be thought of as statistical instructions, and goodness-of-fit can be thought of how well close our model got us to the observations (our target – in our example the market). In our examples the explanation from each friend was a model. Each model was composed of a number of instructions, and depending on the model, they varied in how well they fitted the required task (getting you to the market). In the first scenario the two models got you to the market but differed in the number of instructions. In the second scenario the two models had the same number of instructions but differed in their ability to get you to the market.

14.8.3   Bayesian approaches

Bayesian[10] approaches make use of prior knowledge (known as 'priors'). A prior is then combined with new data through a model to produce updated knowledge (which is called a posterior). A prior can include the results of previous study, or even personal beliefs. Just like information-theoretic approaches, Bayesian approaches can use model selection. But rather than use AIC, different criteria are used (e.g Deviance Information Criteria and Bayesian Information Criterion). Bayesian approaches are particularly useful when you are building upon prior knowledge generated by previous research. The major downside of Bayesian approaches is that they are harder for beginners to understand.

14.8.4   Machine learning

Machine learning refers to computer-centred methods which focus on making accurate prediction rather than trying to understand the relationships between variables. Unlike statistical approaches, machine learning doesn't rely on probability distributions[11]. The simplest forms of machine learning work by creating models that predict the response variable through a set of decision rules based on information contained in the data.

In machine learning, model selection is mostly accomplished through a process called cross-validation. In cross-validation the data is separated into

---

[10]named after Reverend Thomas Bayes who developed the theorem in the 18th century

[11]in this way they are a non-parametric approach

training and testing data sets. Training data sets are used to build models and the testing data sets are used to see how well the models predict the outcomes. Cross-validation is not limited to just machine learning and can be used to compare different modelling approaches in terms of their predictive ability. Some aspects of machine learning and cross-validation are introduced in Chapter 17.

14.9   Under-fitting and over-fitting

There are two types of risk when developing a model:

1. **Under-fitting**: a situation where we have too little complexity (too few parameters) in the model. Consequently, the model has poor goodness-of-fit and consequently is poor at making predictions.

2. **Over-fitting**: a situation where we have too much complexity in the model (too many parameters). As a consequence, these models will describe patterns which don't really exist. An over-fitted model will appear to have good goodness-of-fit with our initial (training) data set, however, it will make poor predictions because it is modelling statistical noise rather the true pattern.

These situations can be summarised in Figure 14.5:

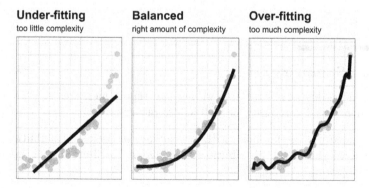

FIGURE 14.5
Model complexity and fit

The use of cross-validation and information criteria are ways of preventing the selection of under-fit and over-fit models. In cross-validation, as training and testing data sets are separated, under-fitted and over-fitted models can often be recognised through the inaccuracy of their predictions. By contrast, approaches using information criterion methods try to prevent under-fitting and over-fitting through equations which take account of goodness-of-fit while penalising the number of parameters.

The advantage of using cross-validation over information criterion methods is it gives an independent measure of a model's predictive performance. The major downside is that the method suits large data sets and can be slow to compute. By contrast methods using information criteria are computationally fast, and suitable for small data sets. The major downside of information criteria is that they cannot give an independent measure of a model's predictive performance by themselves.

14.10   Recommended resources

Print:

- David Anderson (2007) Model based inference in the life sciences: a primer on evidence. Springer Science & Business Media. A readable introduction to information theoretic approaches.

- Michael A. McCarthy (2007) Bayesian methods for ecology. Cambridge University Press. Although written specifically for the WinBugs program, it contains a simple beginners-level introduction for people interested in Bayesian approaches.

14.11   Summary

- When we describe a variable we need to report measures of both central tendency and variability, as well as sample size.

- We need to be familiar with three key statistical distributions: the Bernoulli (binomial) distribution, the Poisson distribution, and the normal distribution.

- There are a variety of modelling approaches. The major ones are:
    - Null hypothesis testing.
    - Information theoretics (which will be the focus of the next chapters).
    - Bayesian inference.
    - Machine learning (which will be introduced in Chapter 17).

15

Understanding linear models

In this chapter we will introduce:

- Graphing model formulas.

- Building models.

- Making predictions.

- Examining alternative models, model outputs, and diagnostics.

15.1 Regression versus classification

In data science predicting a response variable from explanatory variables is known as supervised learning. Within supervised learning there are two major types of analysis:

- Regression: in which we try to predict the value of a response variable based on a set of explanatory variables. For example:

    - Based on income, gender, and family size, can we accurately predict how much protein a person eats each week?

- Classification: in which we try to predict the category of a response variable based on a set of explanatory variables[1]. For example:

    - Based on their survey answers can we accurately predict predict how a person will rate our project (e.g. good, okay, or bad)?

In this chapter we will examine linear models which are a widely used regression technique. By the end of the chapter you should, be able to make and test simple linear models.

---

[1]we will introduce classification later in Chapter 17

15.2   The packages

In this chapter we will make some ggplots using the `tidyverse` package. We will use an information theoretic approach and undertake model selection – for this reason we need the `AICcmodavg` package. Later we will want to simulate dummy variables and for this we will need the `caret` package:

```
library("AICcmodavg")
library("caret")
library("tidyverse")
```

15.3   The data

For this chapter we will use a generic data frame called `mydata`. It is built using an equation for a linear model (which will be explained in detail later). The code for building it is:

```
set.seed(13)
x1 <- rnorm(n = 9, mean = 6, sd = 1.5)
x2 <- c("C","A","C","A","B","B","C","B","A")
error <- rnorm(n = 9, mean = 0, sd = 0.5)

y <- 2 + 1 * x1 + error # the linear equation
mydata <- data.frame(x1, x2, y)
```

The data frame has 3 variables:

1. `x1`: an explanatory variable – which, in this case, is a normally distributed random number.

2. `x2`: an explanatory variable – a categorical variable made up of three different factors: "A", "B", and "C").

3. `y`: our response variable – which is predicted from a linear equation which has an intercept of 2, varies in direct proportion with the value of `x1` (due to a slope of 1), and has an unexplained (normally distributed) error. Note:

in no way does x2 contribute to y although it may look like it from later graphs.

If you don't understand the meaning of the text above don't worry – we cover it in the next couple of sections.

15.4 Graphing a *y* variable

Before we introduce linear models and how to graph them we should be comfortable imagining a single variable and its mean as a graph. This represents the most simple linear model possible. Let's begin with the variable mydata$y:

```
mydata$y

[1]   9.384063   7.032795  10.893681   7.600488   8.785776
[6]   8.403361   9.747286   9.053235   7.502257
```

Whose mean is:

```
mean(mydata$y)

[1] 8.711438
```

We can imagine mapping the individual observations of the variable (y) and its mean to a number line. A number line is simply a graph with one axis. We don't need a special graph for this – it is just the y-axis of a graph. As we are not considering any other variables we don't need to worry about having an x-axis (Figure 15.1). This type of model is usually called an intercept only model.

15.5 What is a linear model?

Linear models predict a continuous response variable in terms of one or more explanatory variables. Linear models are mathematical equations where the

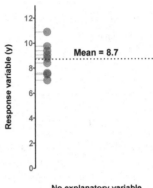

FIGURE 15.1
Graphing a mean (an intercept only model)

response variable can be predicted through the addition of the products of the explanatory variable and their associated slope.

Often the response variable is presented in equations with the letter y. If there is only one explanatory variable it is usually appears as x in equations. If there is more than one explanatory variables, they are usually named: x_1, x_2, x_3 etc. in equations. The slope associated with the explanatory variables usually appears in equations as the Greek letter β, and is numbered with a subscript according to the explanatory variable it is associated with (i.e. β_1, β_2, β_3 etc.)

15.5.1   How to draw a linear model from an equation

When we are graphing a response and explanatory variable we usually plot the response variable on the y-axis, and an explanatory variable on the x-axis. If we wanted people to be able to draw the same line on such a graph we could do so with just two pieces of information as instructions:

1. Knowing where the line cuts the y-axis (which we call the intercept).

2. Knowing the slope of the line.

   The slope can be thought of as the amount of change on the y-axis for a change of 1 unit on the x-axis. A slope with a positive value goes upwards, a slope with a negative value goes downwards (see Figure 15.2).

With these two pieces of information we can draw a straight line on a graph. By extension we can also predict the value of a response variable if we know the equation describes its relationship with the explanatory variable.

The basic equation of a simple linear model is:

$$y_i = \beta_0 + \beta_1 \times x_i + \epsilon$$

This equation may look a little scary but once we know what the parts are, it should become much less daunting:

1. y_i is the predicted value of a continuous response variable based on the observations recorded at row i in our data. For example: y_6 would be the prediction based on the sixth row of our data.

2. β_0 is the value of the intercept.

3. β_1 is the value of the slope.

4. x_i is the value of an explanatory variable, x, at row i in our data set. For example: x_6 would be the observation of x on the sixth row of our data.

5. ϵ is the error term which represents the unexplained variability.

Usually, this equation is written in a slightly simplified form ignoring the error term, and the reference to the data row, and representing the multiplication of β_1 and x as $\beta_1 x$:

$$y = \beta_0 + \beta_1 x$$

In Figure 15.2 we can see that there are 3 lines with the same intercept but different slopes:

1. A upwards line with the equation $y = 5 + 1x$: intercepting the y-axis at 5 with a slope of +1.

2. A horizontal line with the equation $y = 5 + (0 * x)$ which is the same as $y = 5$: intercepting the y-axis at 5 but having a slope of 0.

3. A downwards line with the equation $y = 5 - 1x$: intercepting the y-axis at 5 with a slope of -1.

15.6 Predicting the response variable

If we are given a linear equation we can predict the value of a response variable from the explanatory variables simply by putting their values into the linear

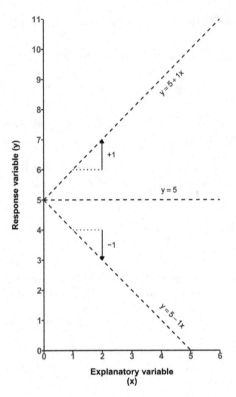

FIGURE 15.2
Example lines and their equations

```
x <- 1:5
x

[1] 1 2 3 4 5
```

We can then use the equation $y = 5 + 1x$ to predict the response variable, y, for each of the values of x:

```
y <- 5 + 1 * x
y

[1]   6   7   8   9  10
```

Which can be plotted on the graph using the **geom_abline()** function[2]:

```
ggplot()+
  geom_point(aes(x = x, y = y), size = 4, alpha = 0.3)+
  geom_abline(intercept = 5, slope = 1)+
  scale_y_continuous(limits = c(0, 11))
```

Which results in:

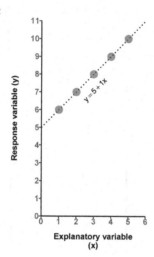

15.7   Formulating hypotheses

A model is the same as a hypothesis – a plausible explanation for some kind of pattern. For any given pattern most people will be able to think of multiple models. We could imagine three simple models for the value of y, our response variable, in mydata e.g.:

1. Model 1 – the value of y has no relationship with either x1 or x2.

2. Model 2 – the value of y is determined by the value of x1.

3. Model 3 – the value of y is determined by the category of x2.

Often it is helpful to visualise these models as graphs (Figure 15.3). We can then imagine a line (our model predictions) which describe the central tendency of each model, for example:

---

[2]because x and y are vectors (not a data frame) we don't need to use the *data* argument in the ggplot

1. Model 1 – A single horizontal line (reflecting a simple mean – often referred to as an intercept only model).

2. Model 2 – A sloping line (with an intercept and slope).

3. Model 3 – Three horizontal lines (with a separate intercept for each of the three categories)

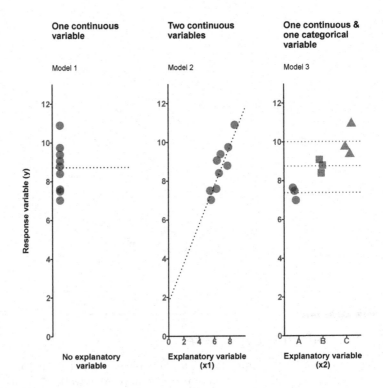

FIGURE 15.3
Visualising three alternate models as graphs

15.8 Goodness-of-fit

Goodness-of-fit answers the question: how well does a model explain the variability seen in the observations? To understand goodness-of-fit it is helpful to understand residuals and correlation.

15.8.1 Residuals

Just by using our eyes we can tell that none of the data in our three graphs (Figure 15.3) matches the model predictions perfectly (the dotted lines). There is always some variation between our observations (data points) and our model predictions. If we zoom in we can visualise this variation as the distance between our observation and the prediction generated by our model. We call the distance between each observation and the model prediction a residual (e.g. for Model 2 – Figure 15.4).

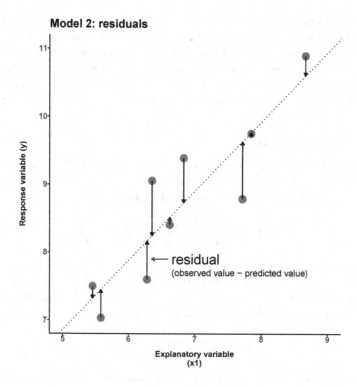

FIGURE 15.4
Residuals (used in the calculation of residual sum of squares)

If all the observations were predicted perfectly by the model there would be no residuals and you would have perfect goodness-of-fit. The greater the residuals the poorer the fit of the model.

If we added all the residuals together their sum would be zero as half are always positive (above the model prediction) and half are always negative (below the model prediction). So to get around this statisticians square each residual[3] and then sum them to get a measure of the overall variation around

---

[3]this is a similar problem that the standard deviation resolves

the model prediction. We call this the residual sum of squares (RSS). The RSS is used to calculate a variety of different goodness-of-fit statistics[4]. Residuals relate to two of the key assumptions of linear models:

1. The residuals should be normally distributed (note: it is the residuals which need to be normally distributed not the response variable itself).

2. The size of residuals should be constant across the range of values predicted by the equation.

Another important statistic is the total sum of squares (TSS) which is the sum of the squared differences between the observed values and the mean of the response variable (Figure 15.5). The TSS is just the RSS of an intercept only model. Given that intercept only models are uninformative (as they don't have any explanatory variables) the TSS can then be used as a baseline for comparing the goodness-of-fit of other models.

15.8.2   Correlation

When variables share some kind of a relationship we we call this a correlation. One measure of the strength of the correlation is known as R-squared. R-squared is usually written as r^2 and is calculated as:

$$r^2 = 1 - \frac{\text{residual sum of squares (RSS)}}{\text{total sum of squares (TSS)}}$$

R-squared is a measure of goodness-of-fit. In a perfect correlation $r^2 = 1$ (meaning the variables track each other perfectly), if there is absolutely no relationship $r^2 = 0$. The R-squared tells us the amount of variation explained by the variables. It can be also be calculated as:

$$r^2 = \frac{\text{Explained variation}}{\text{Total variation}}$$

A model where $r^2 = 0.73$ means the explanatory variables explain 73% of the variability in the model. It also means that 27% of that variability is not explained. We can see in Figure 15.6 that as r^2 increases unexplained variation in a model decreases.

So far we have only considered r^2 for a model with one explanatory variable, however, as we shall see in Chapter 16 we can make models with multiple explanatory variables. We might expect a high r^2 value when we are dealing with a response variable which can be explained by only a few explanatory variables. However, in reality, most response variables are dependent on many explanatory variables[5]. As a result, in some situations, it may be impossible

---

[4]the term has other names and abbreviations including the sum of squared residuals (SSR) and the sum of squared estimate of errors (SSE)

[5]this is why an alternative name for the response variable is the 'dependent variable'

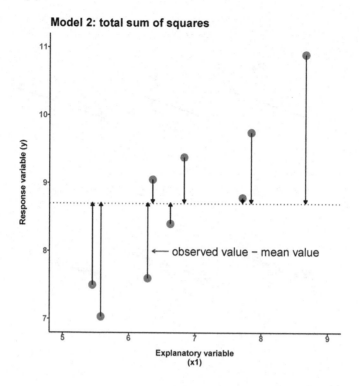

FIGURE 15.5

Differences between observed values and the overall mean (used in the calculation of total sum of squares)

to record enough explanatory variables to achieve a high r^2 value given a project's budget. Under these circumstances even our best model may have a large amount of unexplained variation and, as a result, produce a low r^2 value. Not surprisingly, models in conservation and development may result in r^2 values that would be considered very low by other disciplines (e.g. the physical sciences) due to the large number variables involved (especially for models associated with sociological and ecological problems).

It is also important to remember that although two variables may be closely correlated this does not necessarily mean one causes the other. In some situations the two variables may be correlated due to an unmeasured third variable (e.g. the number of births in a city may be correlated with the number of cars due to population size influencing both). However, other times a strong correlation will exist because of direct causation (e.g. the number of houses in a city and its population).

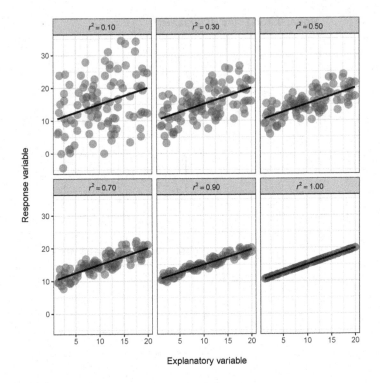

FIGURE 15.6
As r^2 increases unexplained variation in a model decreases

15.9   Making a linear model in R

To make a linear model in R we use the **lm()** function (short for linear model). In this function we have to give the following arguments:

1. A model formula containing:

   (a) Response variable: in the form of $y \sim$ where y is the name of our response variable and a $\sim$ sign (called a tilde) substitutes for an equals sign.

   (b) The names of the explanatory variables in the model. These follow the tilde($\sim$) and complete the model formula e.g. $y \sim x1$, *catch.weight* $\sim$ *hours, cases* $\sim$ *intervention*. If we want an intercept only model we substitute a *1* for the absent explanatory variable i.e. $y \sim 1$.

2. The data e.g. *data = mydata*.

   The basic form of **lm()** looks like:

```
lm(y ~ x1, data = mydata)
```

When this code is run R estimates the line of best fit by minimising the error associated with the RSS. This process is known as ordinary least squares regression.

15.10 Introduction to model selection

In the sections above, we outlined three models which describe different ways for predicting the response variable y. Our question is now which one of these best describes our data? As we are taking an information-theoretic approach we will base our model selection process on Akaike's Information Criterion[6]. The equation for AIC is:

$$AIC = 2K - 2L$$

where

- K = the number of parameters

- L = the log-likelihood of the model

AIC is based on the goodness-of-fit, as measured by log-likelihood (L), with a penalty for the number of parameters (K). AIC can be thought of as a measure of model quality. The lower the AIC value the better the model.

15.10.1 Estimating the number of parameters: K

We can count the number of parameters (which is equivalent to the number of mathematical instructions) by looking at the graphs of our three models. You may recall that each linear model also has an error term associated with it. For the purposes of the calculation of AIC is this also counted as a parameter. As a result, we add one to the count of parameters for each model:

- Model 1: has an intercept only:

  = 1 instruction + 1 error term *(2 parameters)*

- Model 2: an has intercept and slope:

---

[6]see Chapter 14 for a basic description

= 2 instructions + 1 error term *(3 parameters)*

- Model 3: has 3 intercepts:

= 3 instructions + 1 error term *(4 parameters)*

15.10.2   Goodness of fit: L

The measure of goodness-of-fit in AIC is known as the log-likelihood which can be calculated from the residual sum of squares and sample size (see Appendix for step-wise calculations). If there are two models with the same number of parameters then one with the lowest log-likelihood will have the lowest AIC. If there are two models with the same log-likelihood then the model with the fewest parameters will have the lowest AIC.

15.11   Doing model selection in R

To undertake model selection in R first we need to do the following steps:

1. Load a package which handles model selection (such as `AICcmodavg`)

2. Make a list object (to contain our list of candidate models).

3. Assign models to a position in the list.

4. Give each model a simplified name.

5. Run a function to create an AIC or AICc (AIC with a correction for small sample size) table for the models in the list.

   Current advice[7] is to use AICc if the total sample size is smaller than $40\times$ the number of parameters. As AICc converges to AIC when the sample size is large it has been recommended to use AICc over AIC.

Step 1: load a model selection package:

```
library(AICcmodavg)
```

---

[7]Burnham, K. & Anderson, D. (2002) Model Selection and Multimodel Inference: A Practical Information – Theoretic Approach, 2nd edn. Springer, New York

Step 2: make a list object (called in this example `Cand.model`):

```
Cand.model <- list()
```

Lists are another type object class (like data frames and vectors). They are literally a numbered list. The list can be made to contain different classes of objects including vectors, data frames, and even other lists. By using the **list()** function in the code below we can store the model outputs of different linear models in an object.

Step 3: Assign models to a position in the list:

```
Cand.model[[1]] <- lm(y ~ 1, data = mydata)
Cand.model[[2]] <- lm(y ~ x1, data = mydata)
Cand.model[[3]] <- lm(y ~ x2, data = mydata)
```

In the code above, the number in the double brackets `[[ ]]` puts the output of each model into an ordered position within the list. For example, to see the output of the 3rd model we would run `Cand.model[[3]]`.

Step 4: Give each model a simplified name:

```
Modnames <- paste("Mod", 1:length(Cand.model))
```

This line of code gives a short name to each model in the order they appear in the list (e.g. 'Mod 1', 'Mod 2', and 'Mod 3'). The **paste()** function allows us to combine different elements such as text and numbers into a single piece of text (known as a string).

Step 5: Run a function to create an ordered AIC (or AICc table):

```
aictab(cand.set = Cand.model, modnames = Modnames,
       sort = TRUE)
```

By default the **aictab()** function returns an AICc table. If we it to return an AIC table we include the argument *second.ord = FALSE*. If we need to control the number of digits in the output we have to do this outside of the function – by using the **print()** function:

```
aictab(cand.set = Cand.model, modnames = Modnames,
       sort = TRUE) %>%
  print(digits = 3)
```

15.11.1   Interpreting an AIC Table

The code above results in the following table:

```
Model selection based on AICc:

        K   AICc Delta_AICc AICcWt Cum.Wt      LL
Mod 2 3 24.547      0.000  0.856  0.856  -6.874
Mod 3 4 28.214      3.666  0.137  0.993  -5.107
Mod 1 2 34.144      9.597  0.007  1.000 -14.072
```

When you view this table you should notice that it reports an 'AICc' table – this is because it is automatically applying a small sample correction. The table automatically is ordered from the model with the lowest AICc value to the highest. The table has a number of columns, these represent (from left to right):

- The simplified model name (which we gave in Step 4).

- K: The number of regression parameters which includes an additional parameter for the calculation of the error term. This is equivalent to the number of mathematical instructions required to build the model.

- AICc: the value of AICc calculated from the model (which is AIC plus a small sample correction). Lower values represent better models.

- Delta_AICc: the difference in AICc between the each candidate model's AICc and the top model's AICc (sometimes written as ΔAICc). Therefore the Delta_AICc for the top model will always be zero.

- AICcWt: is the AIC weight (often also called 'model weight' or 'model support') which represents the likelihood of the model being the best model of the list given the data. The higher it is the more confidence we should have in the model.

- Cum.Wt: is the accumulative AICcWt (which as it approaches 1 indicates that there is little chance of any lower ranked models having any meaningful support).

- LL: is the log-likelihood. This is equivalent to the goodness-of-fit. Lower values indicate a better fit. If two models have the same log-likelihood that means they have the same goodness-of-fit.

As we can see in our AICc table Model 2 appears to be the model which offers the best explanation of the data because it has approximately 86% of the model weight (AICcWt = 0.856). However, you may notice that Model 3 has a better goodness-of-fit (LL = −5.107) but is ranked lower because it has been penalised for having a larger number of parameters ($K = 4$). Model 1, by comparison, has almost no support (as AICcWt = 0.007).

Did model selection pick the correct top model?
Based on model selection with AICc we have identified that the top model (with approximately 86% model weight) was Model 2. This was indeed the model which we used to simulate the data set (see section 15.3).

15.11.1.1 Evidence ratios

One of the useful features of AICc weight is that it allows us to calculate, how much better one model is compared to another. This is known as an evidence ratio and is simply the model weight of the two models divided by each other. So if we wanted to know how much better Mod 2 was than Mod 3 was we would simply divide the weight the weight of one model by the other e.g.:

```
0.856 / 0.137

[1] 6.248175
```

This reveals Mod 2 is approximately 6× more likely than Mod 3 to be the better model given our current data.

15.11.1.2 Keep in mind

- The top model is just the best of our candidate models. However, our candidate model set may not be a good reflection of the real world.

- It is entirely possible our candidate model set is not very informative. For this reason I recommend including an intercept only model ($y \sim 1$) which can act as a base line for poor model performance (i.e. at a minimum, our models must have a lower AICc than the $y \sim 1$ model).

- While we might want to accept the top model as the best explanation of our data we must investigate any models which share a large proportion of AICcWt as possible alternate explanations.

- We will often come across a situation where the log-likelihood of two models are very close or even the same, but they differ in the number of parameters by one (but otherwise share all the same parameters). In this situation, the model with the higher parameter count has an extra parameter (either a continuous variable or two factor categorical variable) which is not improving the goodness-of-fit (if we are using AIC rather than AICc the model will also have a Delta_AICc which is close to 2 as the penalty term for AIC is 2 × the number of parameters). As a result, we say that this model has a 'pretending parameter'. We need to keep in mind that such a model represents a duplication of an existing model, albeit with an uninformative extra parameter.

- If our sample size is small, model selection might not correctly rank a model we know to be true as better than other models. So, to ensure we have a data set of sufficient size we should always simulate our data and analysis prior to data collection[8]. In this process we should test that our model selection is able to correctly identify the most complicated model as the top model (if it were true) given our sample size. If this test fails our options are to try and increase the sample size or, if that is not possible, design models with fewer parameters.

- It is also possible to combine results from several models together in a process called model-averaging, which uses the model weights, to improve the estimation of the regression coefficients (not covered in this book).

15.12   Understanding coefficients

Let's imagine we are interested in the outputs of all three models we recently graphed. We will want to know the estimated values of the coefficients in (the value of the βs) in order to establish the predictive equations and graph the models. These coefficients can be extracted in a variety of ways in R. For simplicity, we will use the **coef()** function to view these coefficients.

For **Model 1**, lm(y ~1), there is just 1 value because there is only 1 regression parameter: an intercept (i.e. the mean):

---

[8]code for simulation is given in section 15.18

```
coef(Cand.model[[1]])

(Intercept)
   8.711438
```

For **Model 2**, lm(y~x1), there are 2 values because there are 2 regression parameters: the intercept and slope of x1:

```
coef(Cand.model[[2]])

(Intercept)          x1
   1.751748    1.021096
```

For **Model 3**, lm(y~x2), there are 3 values because there are 3 regression parameters: an intercept (the mean of x2 for factor 'A'), and coefficients corresponding to the other two categories (x2 for factor 'B') and (x2 for factor 'C'):

```
coef(Cand.model[[3]])

(Intercept)          x2B          x2C
   7.378513    1.368944     2.629830
```

Examining model structure

All objects created in R have a structure. A model which is saved as an object will store key information that relates to that analysis. We can investigate what information has been stored with the **str()** function:

```
str(Cand.model[[3]])
```

If we run this code we can see the first element is called $ coefficients. By adding this element to our model we can extract the coefficients and get the same output as the **coef()** function:

```
Cand.model[[3]]$coefficients

(Intercept)          x2B          x2C
   7.378513    1.368944     2.629830
```

15.13   Model equations and prediction

You will recall that the basic equation of a line is:

$$y_i = \beta_0 + \beta_1 x_i$$

By substituting each of our coefficients for the matching β terms we can predict the value of y for each observation (i). Let's attempt this for the first row of our data which is given by:

```
mydata[1, ]

      x1 x2         y
1 6.83149   C 9.384063
```

Model 1:

$$y_i = intercept$$

therefore

$$y_1 = 8.71$$

Model 2:

$$y_i = intercept + \text{slope } x_i$$

where

$$y_i = 1.75 + 1.02 x_i$$

therefore

$$y_1 = 1.75 + 1.02 \times 6.83 = 8.72$$

Model 3:

$$y_i = intercept + \text{slope } x_2(B) + \text{slope } x_2(C)$$

where

$$y_i = 7.38 + 1.37 x_2(B) + 2.63 x_2(C)$$

therefore

$$y_1 = 7.38 + 1.37 \times 0 + 2.63 \times 1 = 10.01$$

You will notice that none of the models predicted the actual value of y_1 (9.38) particularly well. This is not unexpected. We need to keep in mind that unless there is perfect goodness of fit (i.e. $r^2 = 1$) our predicted line of best fit will only ever match the observations imperfectly.

Did model selection predict the correct model equation?

Based on our model coefficients we see that the predictive equation of Model 2 was:

$$y = 1.75 + 1.02x.$$

Thanks to our simulated data we can compare our predictive equation to the code we used to create the data set:

```
y <- 2 + 1 * x1 + error
```

which is the same as:

$$y = 2 + 1x.$$

So, it appears not only did our model selection pick the correct top model but our predictive equation came very close to predicting the values which created the data.

15.13.1 Dummy variables and a design matrix

The equation for Model 3 may appear strange because at first glance it appears to have an extra coefficient for slope. This is a result of x_2 being a factor with three levels. Mathematically, each of these three factors can be described using two columns of data: which relate to the factors 'B', and 'C'. Each of the cells in these columns can only have a value of either 0 or 1 (these are called dummy variables). As a result, x_2 is equivalent to having two parameters describing B-ness and C-ness. Our factor 'A' is just a case where both 'B' and 'C' are zero. Together all these dummy columns represent a plan for predicting our response variable. By adding a dummy column for our intercept (which gives a 1 to every row in the data frame) we have created a design matrix (Table 15.1) for our model. A design matrix shows us how to combine the coefficients to make prediction for every row of our data set.

To predict our response variable for Model 3 we multiply the dummy variables by our coefficients, and sum the products. As a result we can make a prediction for each of the x_2 factors from Model 3:

For $x_2(A)$:
$$y_1 = 7.38 + 1.37 \times 0 + 2.63 \times 0 = 7.38$$

For $x_2(B)$:
$$y_1 = 7.38 + 1.37 \times 1 + 2.63 \times 0 = 8.75$$

For $x_2(C)$:
$$y_1 = 7.38 + 1.37 \times 0 + 2.63 \times 1 = 10.01$$

We can also use the **model.matrix()** function to produce the design matrix for each row of our data automatically:

TABLE 15.1

The design matrix for a model with a categorical variable (Model 3)

	Intercept	x_2(B)	x_2(C)
y = intercept + x_2(A)	1	0	0
y = intercept + x_2(B)	1	1	0
y = intercept + x_2(C)	1	0	1

```
model.matrix(Cand.model[[3]])

  (Intercept) x2B x2C
1           1   0   1
2           1   0   0
3           1   0   1
4           1   0   0
5           1   1   0
6           1   1   0
7           1   0   1
8           1   1   0
9           1   0   0
attr(,"assign")
[1] 0 1 1
attr(,"contrasts")
attr(,"contrasts")$x2
[1] "contr.treatment"
```

15.13.2   Plotting a prediction with geom_abline()

We can graph our predictions on a ggplot by adding a layer with the **geom_abline()** function, and inserting our coefficients into the *intercept* and *slope* arguments:

Model 1:

```
ggplot()+
  geom_point(data = mydata, aes(x = 1, y = y))+
  geom_abline(intercept = 8.71, slope = 0)
```

Model 2:

```
ggplot()+
  geom_point(data = mydata, aes(x = x1, y = y))+
  geom_abline(intercept = 1.75, slope = 1.02)
```

Model 3:

We use **geom_abline()** somewhat differently for Model 3. This is because when we use dummy variables the equation results in three predictive lines (one for each category): A (the intercept), B (the intercept + offset for B) and C (the intercept + offset for C). As a result, we need a separate **geom_abline()** for each category:

```
ggplot()+
  geom_point(data = mydata, aes(x = x2, y = y))+
  geom_abline(intercept = 7.38, slope = 0) +         # x2 = 'A'
  geom_abline(intercept = 7.38 + 1.38, slope = 0) + # x2 = 'B'
  geom_abline(intercept = 7.38 + 2.63, slope = 0)   # x2 = 'C'
```

15.13.3 Automatic prediction

Rather than calculate the predicted values manually we can use the **fitted()** function to make the predictions. The predictions for our top model, Model 2, are shown below (note: these values will differ slightly from manual estimates due to rounding):

```
fitted(Cand.model[[2]])

       1         2          3          4          5          6
8.727358  7.449050  10.597246   8.165234   9.628270   8.514765
```

```
      7         8         9
9.761493  8.240836  7.318690
```

15.14   Understanding a model summary

There is more to a model than just coefficients. The **summary()** function produces a more comprehensive model output:

```
summary(Cand.model[[2]])
```

```
Call:
lm(formula = y ~ x1, data = mydata)

Residuals:
    Min       1Q    Median       3Q       Max
-0.84249  -0.41625  -0.01421   0.29643   0.81240

Coefficients:
            Estimate Std. Error t value Pr(>|t|)
(Intercept)   1.7517     1.3378   1.309  0.23173
x1            1.0211     0.1942   5.259  0.00117 **
---
Signif. codes:
0 '***' 0.001 '**' 0.01 '*' 0.05 '.' 0.1 ' ' 1

Residual standard error: 0.5889 on 7 degrees of freedom
Multiple R-squared:  0.798,        Adjusted R-squared:  0.7692
F-statistic: 27.66 on 1 and 7 DF,  p-value: 0.001174
```

The output has the following elements:

- **Call:** the model formula

- **Residuals:** gives the a summary of the residuals in quartiles (25% percentile intervals).

- **Coefficients:** a table with 5 components:

  1. the name of the coefficients (i.e. the regression parameters).

  2. **Estimate**: the estimated values of the model parameters. The result is a rounded version of that given by the **coef()** function.

  3. **Std. Error**: the standard error (the standard deviation) of our coefficient estimates.

  4. *t* **value**: a measure of how many standard deviations each coefficient estimate is from zero.

  5. **Pr($>$|t|)**: the *p* value of the estimate. Effectively this is a test of whether the estimate could actually have a true value of zero. Lower Pr($>$|t|) values indicate that this is less likely, and are marked with symbols outlined in the 'Signif. codes' output line (see below). This is a hold over from null hypothesis testing approaches. A significant *p* value will result when if the observed *t* value exceeds the critical *t* value (which is dependent on the degrees of freedom of the model).

- **Signif. codes:** the key for the symbols which relate to the approximate value of **Pr($>$|t|)** .

- **Residual standard error:** an overall measure of how much the observations differ from the predicted model. The degrees of freedom (of the residuals) quoted are number of the data points that were used to determine the coefficients (number of observations minus the number of coefficients).

- **Multiple R-squared:** the proportion of variation explained by the model. This value increases as number of explanatory variables increases – as a result, the Adjusted R-squared is preferred as a measurement.

- **Adjusted R-squared:** the proportion of variation explained by the model adjusted by the number of explanatory variables (unlike r^2 where the range is between 0 and 1, the adjusted R-squared can result in slightly negative values). Prior to the development of information criteria this was one of the primary tools for measuring goodness-of-fit. In our example, the model explains approximately 77% of the variability in the data which means that around 23% of the data remains unexplained by the model.

- **F-statistic:** is a test statistic, measuring whether our linear regression model provided a better fit to the data than a intercept only model (i.e. y~1). It is presented with the degrees of freedom which allow the *p* value to be estimated from the *F* statistic[9] (which requires two measurements for its degrees of freedom: the number of coefficients minus 1, and the number of observations minus the number of coefficients).

---

[9]an equivalent *p* value will be produced by `anova(Cand.model[[1]], Cand.model[[2]])`

What are degrees of freedom?
Degrees of freedom represent the amount of independent information used to
calculate the coefficients of a model. If we add more coefficients to the model
without increasing the sample size the amount of information we have to
estimate the coefficients decreases, and as a result our coefficients will become
less precise.

15.15   Standard errors and confidence intervals

Just as we have the standard deviation for measuring the deviation around a
mean, we have a similar statistic for measuring the deviation around the estimated values associated with our model parameters (the coefficients). We call
this statistic the standard error. The standard error forms the foundation for
building 'confidence intervals' around these model parameters. The size of the
confidence intervals reflects the precision of our estimates (smaller being more
precise, larger being less precise). Confidence intervals can be built around
any percentage value (such as 90% or 99%) but most commonly 95% is used.
As a result, these are called '95% confidence intervals'. 95% of the time the
true value of the parameter (of the population) will fall within the bounds
given by a 95% confidence interval. However, we can never be sure that our
confidence interval actually contains this value.

   To estimate the confidence intervals for our parameters we use the **con-
fint()** function. Additionally, we can use the argument *level* to set the percentage confidence interval (as a proportion). For example, a 95% confidence
interval will be given by:

```
confint(Cand.model[[2]], level = 0.95)

                2.5 %    97.5 %
(Intercept) -1.4116811 4.915178
x1           0.5619945 1.480198
```

15.15.1   Confidence intervals for model predictions

How do we know if our model prediction is reliable? If we were to collect a
similar set of data again would we get similar model predictions? Again we
can assess this situation by using confidence intervals. However, this time the

confidence interval is based around the residual standard error rather than the standard error. Again, the size of the confidence intervals reflects the precision of our model. A small confidence interval will suggest that our model should make reliable (precise) predictions, a large confidence interval will suggest that our model may make unreliable predictions.

While the 95% confidence interval describes the uncertainty around our predicted regression line, it does not predict observations. For this reason it is perfectly normal for observations to appear outside the bounds of the confidence interval.

We can automatically plot the 95% confidence interval for linear models with only one explanatory variable using the *method = "lm"* argument within the **geom_smooth()** function:

```
ggplot()+
  geom_point(data = mydata, aes(y = y, x = x1))+
  geom_smooth(data = mydata, aes(y = y, x = x1), method = "lm")
```

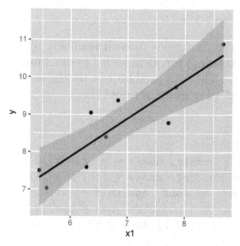

The downside of relying on the **geom_smooth()** function is that it can only produce confidence intervals for the simplest models – those involving one explanatory variable. Alternatively, we can make confidence intervals by using the **predict()** function which can also calculate confidence intervals for complex models:

```
conf <- predict(Cand.model[[2]],
                interval = 'confidence')
```

Which results in the following output:

```
head(conf)

        fit      lwr       upr
1  8.727358 8.263145  9.191572
2  7.449050 6.715836  8.182263
3 10.597246 9.630622 11.563869
4  8.165234 7.640112  8.690357
5  9.628270 9.007488 10.249053
6  8.514765 8.042258  8.987271
```

We can then bind the output to our original data frame with the **cbind()** function for graphing:

```
my.conf <- cbind(mydata, conf)
```

We can then add the confidence interval manually to a ggplot through the use of the **geom_ribbon()** function, with a **geom_line()** to layer to produce the line of best fit. This will produce exactly the same graph that was automatically produced using **geom_smooth()**:

```
ggplot()+
    geom_point(data = mydata, aes(y = y, x = x1)) +
    geom_line(data = my.conf, aes(y = fit, x = x1))+
    geom_ribbon(data = my.conf, aes(ymin = lwr,
                                    ymax = upr, x = x1),
            fill = "grey", alpha = 0.5)
```

15.16   Model diagnostics

After producing a model we should check that it is fulfilling the key assumptions of a linear model. Base R automatically produces a plot of four key model diagnostics using the **plot()** function when it is used in combination with a linear model. The four assumptions are:

1. **Assumption: Data is linear**
   (The data can be described by a linear model)

   (a) Graph: *Residuals vs Fitted*

   (b) Visual test: If the red trend line runs more or less horizontally and the data points are distributed across the graph without a distinct pattern the assumption is okay. A failure would be indicated by the data points following distinct curves. If there is a curve it is likely to be caused by a non-linear explanatory variable. Note: this plot will also pick up issues associated with unequal variability (known as heteroskedasticity) in residuals (which will usually be more evident in the Scale–Location graph).

   (c) Action: Check that the model formula is not directly modified by exponents e.g. `lm(y~x^2)`. If variables need to be transformed this can be done within the data frame, or by the use of special R functions within the equation e.g. **poly()**.

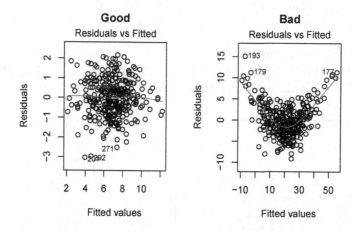

2. **Assumption: Normality of residuals**
   (Residuals are normally distributed)

   (a) Graph: *Normal Q-Q plot*

   (b) Visual test: If the majority of the points fall more-or-less on the dotted line the assumption is okay.

   (c) Action: Trial data transformation (e.g. log transformation), or the use of a non-parametric modelling technique (as they don't rely on statistical distributions).

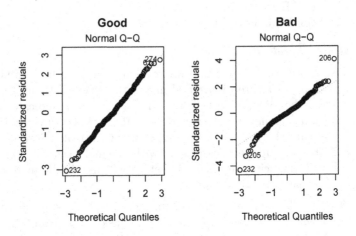

3. Assumption: Homogeneity of residuals
(The size of the residuals is constant regardless of the predicted (fitted) value)

    (a) Graph: *Scale-Location*

    (b) Visual test: If the red trend line runs more or less horizontally and data points are spread points equally across the graph the assumption is okay.

    (c) Action: Trial data transformation (e.g. log transformation)

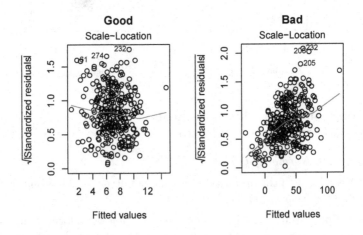

4. Assumption: Outliers are not influential
(Outliers do not have a disproportionate effect on the model)

    (a) Graph: *Residual vs Leverage*

(b) Visual test: Outliers which cross the '1' dotted contour line (Cook's distance) need to be checked, and the effect of their removal tested. Outliers which cross the '0.5' dotted contour line should be checked. Such outliers influence the regression through unusual combinations of response variable values and explanatory variable values. Regardless of their influence the 3 most influential observations are always identified. Influential observations will appear in the top right and bottom right corners of the graph.

(c) Action: Rerun the model with influential data points removed.

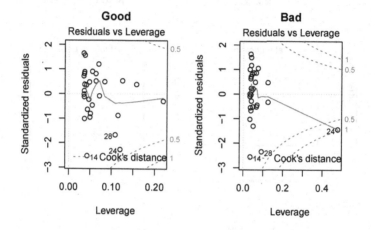

In order to view all 4 plots at once we need to partition (split) the R plot window using the **par()** function into 2 rows and 2 columns through the **mfrow()** function[10] (note: these partitioning functions don't work for ggplot). We then use the **plot()** function to show the built-in model diagnostic plots. Diagnostic plots involving models with small sample sizes (like the one used in the example above) will often show some 'bumpiness' in the plots. This is not cause for concern. Indeed, if we increase our sample size this bumpiness disappears.

```
par(mfrow = c(2, 2))
plot(Cand.model[[2]])
```

---

[10]if we need to undo this partition we use: `par(mfrow=c(1,1))`

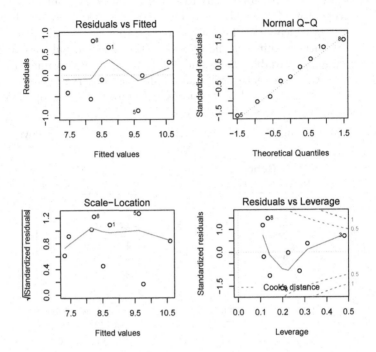

15.16.1   Still problems?

What happens if your model diagnostics, despite your best efforts, still show problems? Usually the first thing to do is ask yourself: am I performing the correct type of test? The R diagnostic plots associated with the **lm()** function assume that the distribution of the residuals will be normally distributed. If your response variable has a binomial or Poisson distribution the diagnostic plots associated with a model made with **lm()** will not be appropriate. In such cases you need to use a different type of regression method (see Chapter 16 for the most common types).

15.17   Log transformations

Linear modelling is based on having residuals which are symmetrical about our regression line. If we are trying to work with skewed data this assumption is violated, and consequently the outputs of such a linear model could be very wrong. Many issues around skewness, and non-normality of residuals can be corrected if the response variable is log transformed (provided the new diagnostic plots show the issues are then resolved). We can see the effect of log transformation on skewness in the graphs below:

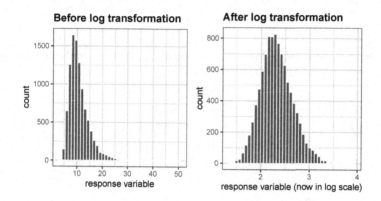

But before we can use such a transformation we should understand what logarithms are.

15.17.1 What are logarithms?

If we think of the number 1,000 we can also describe it in terms of the number 10:

$$10 \times 10 \times 10 \times = 1000$$

We can write this in mathematical notation as 10 to the power of 3 (which is also known as the exponent):

$$10^3 = 1000$$

Which is written in R as:

```
10 ^ 3

[1] 1000
```

In our example the number ten is known as the base. Using the **log()** function in R we can reverse the operation to find out the exponent:

```
log(1000, base = 10)
```

```
[1] 3
```

If we want to express a logarithm in terms of base 10 we can use the **log10()** function as a short cut:

```
log10(1000)
```

```
[1] 3
```

This equivalent to saying that 'the logarithm of 1000 in the base of 10 is 3'. Which can be written in mathematical notation as:

$$\log_{10}(1000) = 3 \qquad\qquad (15.1)$$

Which follows the pattern:

$$\log_{base}(a) = b$$

When

$$base^b = a$$

While it may seem sensible to use a base of 10, the default base of the **log()** function in R is Euler's number[11] (which is written as e) which is approximately 2.718. As a consequence, the **log()** function of 1000 in R is:

```
log(1000)
```

```
[1] 6.907755
```

Which is equivalent to:

$$\log_e(1000) = 6.907755$$

---

[11]the reason behind this preference in R is because e has a number of properties which simplify mathematical equations. This logarithm is widely known as the natural logarithm and is sometimes written as 'ln' in equations

To reverse the operation of the default **log()** function we need to use the **exp()** function.

```
my.exp <- log(1000)
exp(my.exp)

[1] 1000
```

Importantly, even if you can interpret log values easily, most people you will have to communicate with can't. Consequently, it is normally necessary to back-transform our predictions to the number scale people are familiar with.

Understanding growth rates

Logarithms are particularly important for understanding growth rates. For example, in March 2020, in some countries, the number or people infected with the SARS-CoV-2 virus was growing at a rate of ~25% per day. This equivalent to the infected population growing by 1.25× each consecutive day. If we began with just 1 person we can predict the total number infected within the first month:

```
number.infected  <- 1
days <- 1:31
total.infected <-  number.infected * 1.25 ^ days
```

If we graph these numbers we can see the alarming rate of spread of the disease:

```
ggplot()+
  geom_point(aes(x = days, y = total.infected))
```

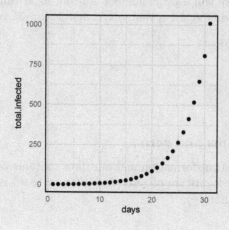

Unfortunately, we can't see the changes happening in the first two weeks because of our scale. By changing the scale to a log scale we can see what would be happening in the first few weeks more clearly. If we want, we can specify the y-axis breaks to help people understand the graph:

```
ggplot()+
  geom_point(aes(x = days, y = total.infected))+
  scale_y_log10(breaks = c(10, 25, 100, 250, 500, 1000, 2000))
```

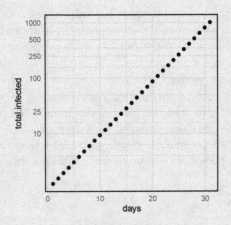

Often when working in community settings it is helpful to talk about the time a human population will take to double given a fixed growth rate. We can solve this with logarithms. For example, if the population of a village is growing at 2% per year how many years will the village take to double in size? Again in order to make our base we change the growth rate to a multiple (i.e. 2% becomes 1.02). As we are wanting to know how long it will take a population to double we will use the number 2 as our logarithm. This reveals that a population growing at 2% a year would double in ~35 years:

```
log(2, base = 1.02)
```

```
[1] 35.00279
```

15.17.2   Logarithms and zero

If we need to do a log transformation and our data contains zeros we have a problem – because the logarithm of zero (regardless of base) is undefined (in R it returns negative infinity):

```
log(0)
```

```
[1] -Inf
```

The way we overcome this by adding a 1 to the variable we are transforming. The easiest way to do this is to use the **log1p()** function to automatically do the transformation. After running the model we can then use **expm1()** function to back transform the predictions back to the normal numerical scale:

```
trans <- log1p(0)
expm1(trans)
```

```
[1] 0
```

15.18 Simulation

To complete our understanding we need to be able to simulate the entire model selection process. Ordinarily we would aim to do this at the Plan phase of the Deming cycle (see Chapter 3). What we want to is ensure that we are collecting enough samples to ensure that our most complex model would come out as the top-ranked model if it were true.

The data we have used so far in this chapter was the result of a single simulation. Our data was the product of luck, and consequently our model selection was also, to some extent, the consequence of luck. The question is: if were to run the same process thousands of times would our top-ranked model still come out as the best? We can answer this question by running a simulation using a for loop with the following steps:

Step 1: Make a reproducible example (with randomly drawn data typical of the values we are expecting).

Step 2: Undertake model selection.

Step 3: Make a loop to repeat the process thousands of times.

Step 4: Check the frequency at which the true model comes out as the top-ranked model.

We have already covered steps 1–2 previously in Chapter 5. In the next sections we cover steps 3–4.

15.18.1 Making a for() loop

To make a loop in R we use the **for()** function. In the function we:

1. Create an vector, i, to act as an index to keep count.

2. Tell the index the values it will have to loop through e.g. i in 1:3 (this will make the process repeat 3×).

3. After the **for()** function enclose the piece of code we want repeated within {}.

A simple example is below:

```
for(i in 1:3) {
  print("My loop")
}

[1] "My loop"
[1] "My loop"
[1] "My loop"
```

However, in order to store the information we need to create an empty vector using (<- NULL) and then get the **for()** loop to store the information in the vector (using the index i.e. [i]). Finally, we have to call the vector to see what information it contains:

```
my.output <- NULL
for(i in 1:3) {
  my.output[i] <- "My loop"
}
my.output

[1] "My loop" "My loop" "My loop"
```

15.18.2 Example simulation

For this example we will run 10,000 simulations – each time generating a new data set and undertaking an analysis – and recording the top-ranked model in the process (this will take a minute or so for R to do these calculations). Notice that we are using the **sample()** function to randomly assign factors rather than giving a predefined order. As we are only recording the top-ranked model name into our object, best.model, we can use the **table()** function to reveal how frequently each model is ranked as the top-model:

```
set.seed(31)

# make empty vector to store our results
best.model <- NULL

# begin the for loop
for (i in 1:10000) {

    # make randomised data set
    x1 <- rnorm(n = 9, mean = 6, sd = 1.5)
    x2 <- sample(rep(c("A","B","C"), 3), size = 9,
                 replace = FALSE)
    error <- rnorm(n = 9, mean = 0, sd = 0.5)

    # equation
    y <- x1 + error

    # make data frame
    mydata <- data.frame(x1, x2, y)

    ## Model selection process
    # Make a list object
    Cand.model <- list()

    # Assign models to a position in the list
    Cand.model[[1]] <- lm(y ~ 1, data = mydata)
    Cand.model[[2]] <- lm(y ~ x1, data = mydata)
    Cand.model[[3]] <- lm(y ~ x2, data = mydata)

    # Give each model a name
    Modnames <- paste("Mod", 1:length(Cand.model), sep = " ")

    # Make AIC table
    our.tab <- aictab(cand.set = Cand.model, modnames = Modnames,
                      sort = TRUE)

    # Store top-ranked model
    best.model[i] <- our.tab$Modnames[1]

    # end loop
}

# Show table of results
table(best.model)
```

Which results in:

```
best.model
   1    2    3
  22 9974    4
```

Our output clearly shows that Model 2 will be the top-ranked model (~99.7% of the time). As a result, we can have high confidence that our sample size will give us a reliable result if Model 2 is the true model. However, to be a fair analysis we should also repeat the process with the most complicated model (Model 3) being simulated as the correct model. This requires us to make y based on the Model 3 design matrix. As Model 3 is based on categorical variables we need to make a set of dummy variables. We can make these using the **dummyVars()** function from the **caret** package using " ~ . " as the formula. We then use the **predict()** function to change the output back to a data frame. Consequently, we could substitute the code below in the simulation (using 1.4 and 2.6 as our assumed coefficients). However, because our code is somewhat more complicated it will take a little longer to complete. **Note:** if you are using an R version ≥ 4 you may have to remove the dot from the names of the dummy variables for the code to run i.e. x2.B becomes x2B:

```r
# make randomised data set
x1 <- rnorm(n = 9, mean = 6, sd = 1.5)
x2 <- sample(rep(c("A","B","C"), 3), size = 9,
             replace = FALSE)
error <- rnorm(n = 9, mean = 0, sd = 0.5)

# make temporary data frame with explanatory variables
df <- data.frame(x1, x2)

# use caret package to make dummy variables
dummy <- dummyVars(" ~ .", data = df, fullRank = TRUE)
dummy.df <- data.frame(predict(dummy, newdata = df))

# get dummy columns
x2.B <- dummy.df$x2.B
x2.C <- dummy.df$x2.C

# equation
y <- 6 + 1.4 * x2.B - 2.6 * x2.C + error

# make data frame
mydata <- data.frame(x1, x2, y)
```

15.19   Reporting modelling results

Project reporting in conservation and development is usually non-technical. Consequently, the major findings should be briefly given in plain language (as an activity summary – see Chapter 18). Detailed statistical results (e.g. modelling) are not usually presented in the main body of the project reports, although it may be appropriate to include a key graph. Rather these, more detailed, results may be included in a concise but readable form as attachments that accompany the report (often called annexes or appendices).

In order for a model selection process to be interpretable by others we need to present the AICc table (or equivalent) with the model formulas in our attachment. The text accompanying the AICc table should describe the top-ranked model(s) and the amount of support (AICc weight) they receive. Given that most people reading the report will be unfamiliar with such tables it is helpful to explain the differences between differently ranked models in terms of their evidence ratios.

Importantly, the consequences of the highest ranked models should be briefly discussed in the context of the project (unless one model has all, or almost all of the model weight). Model selection does not guarantee that the top-ranked model will be particularly informative (rather, it is likely to more informative than other the candidate models). For this reason it is often helpful to present the adjusted r^2 values to give some indication of the goodness-of-fit of the top-ranked models. Given that the main reason we use data science in conservation and development is to develop and improve interventions we are usually interested in understanding the predictive equation made by our modelling. As a consequence, we need to present the model equation and a table of coefficients which shows the uncertainty in our estimates.

Example

In the following example, we give the results from our made-up data set which has variables with uninformative names: y, x_1, x_2. In a real world situation we would use names for variables that are instantly understandable and relate directly to the data (e.g. 'village.income', 'distance.to.market', 'crop' etc.). Details of the modelling should lead by describing the highest ranked model(s) in meaningful terms e.g. 'We found that differences in village income were best explained by the distance people had to travel to market', before describing the outputs of the modelling in detail.

Our modelling suggests that y appears to be best explained by x_1. Our top-ranked model was $y{\sim}x_1$ which had 86% of model support (see table below). By comparison our second ranked model $y{\sim}x_2$ had only 14% of model support. Based on evidence ratios $y{\sim}x_1$ is approximately 6$\times$ more likely to be the top-ranked model than $y \sim x_2$. By comparison there was almost no support for our third model $y \sim 1$.

Model	K	AICc	Delta AICc	AICc weight	log-Likelihood
$\sim x_1$	3	24.55	0.00	0.856	-6.87
$\sim x_2$	4	28.21	3.67	0.137	-5.11
~ 1	2	34.14	9.60	0.007	-14.07

Our top-ranked model predicts y based solely on the basis of x_1 and explains ~77% of data variability[12] with the following equation:

$$y = 1.75 + 1.02x_1$$

The variability associated with the model coefficients is given in the table below:

	Estimate	Std. Error	t value	Pr($>|t|$)
(Intercept)	1.7517	1.3378	1.31	0.2317
x_1	1.0211	0.1942	5.26	0.0012

15.20   Summary

- We use linear models when the response variable is continuous and we believe the model's residuals will be normally distributed.

- Predictive equations can be built from the model's coefficients.

- We can check that the assumptions of linear modelling are met by using R's diagnostic plots.

- Model selection is a process by which we can identify which of our models is (or are) the best explanation.

- Ideally, we want to simulate our data before our study to ensure our sample size is sufficient to adequately test our hypotheses.

---

[12]based on the adjusted R-squared value given in the model summary

16

Extensions to linear models

In this chapter we will examine:

- How to visualise linear models with more than one explanatory variable.

- How generalised linear models extend standard linear models for different distributions.

- How many classical statistical tests are just types of linear model.

16.1 Building upon linear models

Linear modelling, which we introduced in the previous chapter, forms the basis for other types of modelling which can involve many explanatory variables, and different statistical distributions. Indeed many specialised statistical tests are just a type of linear model. In this chapter we shall look at some of the extensions to linear modelling and their underlying similarities.

16.2 The packages

In this chapter we will require packages for model selection (AICcmodavg), checking multicolinearity and generating 'type 3' p values (car), model diagnostics (DHARMa), getting details of models (insight), doing repeated measures analyses (nlme), modelling with ordered categories (ordinal), and graphing (tidyverse). Ideally, these packages should be loaded now:

```
library("AICcmodavg")
library("car")
library("DHARMa")
library("insight")
```

```
library("nlme")
library("ordinal")
library("tidyverse")
```

16.3 The data

In this chapter we will be demonstrating a variety of modelling techniques consequently we will need a variety of data sets:

```
library("condev")
data(attendees)
data(demo1)
data(hunt)
data(prosecutions)
data(traps)
```

16.4 Multiple regression

In the previous chapter we examined linear models with just one explanatory variable. However, we can easily extend linear models to involve multiple explanatory variables – in an approach called multiple regression. Such an approach is important as most real world situations in conservation and development involve multiple explanatory variables e.g.:

- If we are investigating the weight of game animals harvested by hunters we might suspect that this is related to the number of hours the hunters spent hunting and the distance they live from the forest.

- If we are interested in understanding historic assault rates in a region we might suspect that this will be related to regional unemployment rates, average income, population density, and educational opportunities.

In order to be become confident modelling using multiple regression we need to be able to visualise models built from formulas with two explanatory variables. As our models become more complicated (involving more variables) we loose the ability to visualise them as two dimensional graphs. Our understanding of model formulas, however, should allow us to continue build reasonable models even though they may be too complicated to easily visualise. But before we can start visualising multiple regression there are two main types of model structure that we need to understand: additive models and interaction models.

16.4.1   Additive versus interaction models

In additive models the response variable can be calculated by adding together the effects of the explanatory variables. Therefore the equation for an additive model with two explanatory variables would be:

$$y = \beta_0 + \beta_1 x_1 + \beta_2 x_2$$

with the model formula being:

```
lm(y ~ x1 + x2)
```

By comparison, in an interaction model the response variable is calculated by adding together the individual effects of the explanatory variables plus an effect of the explanatory variables interacting together (modelled by multiplying them together). Therefore the standard equation for an interaction model with two continuous variables would be:

$$y = \beta_0 + \beta_1 x_1 + \beta_2 x_2 + \beta_3 x_1 x_2$$

with the model formula being:

```
lm(y ~ x1 * x2)
```

which can also be written as:

```
lm(y ~ x1 + x2 + x1:x2)
```

We could modify this formula to make a model which assumes there is a shared intercept (rather than a separate intercept for each of the explanatory variables):

$$y = \beta_0 + \beta_1 x_1 x_2$$

with the model formula being:

```
lm(y ~ x1 : x2)
```

16.4.2   Visualising multiple regression

The differences between additive and interaction models are easiest to see when we compare different models made up of one continuous variable and one categorical variable (with two categories). In Figure 16.1 we see three types of models: an additive model (A), an interaction model (B), and an interaction model with a shared intercept (C).

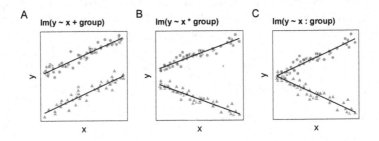

FIGURE 16.1
Multiple regression with two explanatory variables: one continuous, and one categorical variable

Each of these models can be applied to our **demo1** data set in which we have two explanatory variables x (a numeric variable), and **group2** (a factor with two categories: **treat1** and **treat2**). The resulting coefficients give the β value of each of the regression parameters:

- An additive model, lm(y~x+group), Figure 16.1A:

  - 3 regression parameters: intercept, slope, and an intercept offset for one of the categories e.g.:

    ```
    lm(y ~ x + group2, data = demo1)

    Call:
    lm(formula = y ~ x + group2, data = demo1)

    Coefficients:
    (Intercept)              x  group2treat2
         1.1936         0.8660        0.1667
    ```

- An interaction model, lm(y~x*group), Figure 16.1B:

  - 4 regression parameters: intercept, slope, an intercept offset for one of the categories, and a slope offset for one of the categories e.g.:

```
lm(y ~ x * group2, data = demo1)

Call:
lm(formula = y ~ x * group2, data = demo1)


Coefficients:
  (Intercept)                      x      group2treat2
       0.3547                 0.9576            2.6462
x:group2treat2
      -0.2681
```

- An interaction model with a shared intercept, lm(y~x:group), Figure 16.1C:

  - 3 regression parameters: intercept, and a separate slope for each of the 2 categories e.g.:

```
lm(y ~ x : group2, data = demo1)

Call:
lm(formula = y ~ x:group2, data = demo1)

Coefficients:
  (Intercept)  x:group2treat1  x:group2treat2
       1.2435          0.8615          0.8773
```

Note: You may have noticed that in the example above the two different slopes we calculated are essentially the same – so this model is probably not a good explanation of the data. It is important to remember that just because the algorithm we are using finding a solution this is not an endorsement that the model's solution is a good explanation of the data.

16.4.2.1 Visualising continuous variables

While it is easy to visualise differences between additive and interaction models made up of a continuous and categorical explanatory variable, it is harder for two continuous variables. This is because when have two continuous explanatory variables we need an extra dimension to visualise this relationship.

We cannot use separate graphs of $y \sim x_1$ and $y \sim x_2$ to understand the relationship $y \sim x_1 + x_2$ because these graphs will not show that each pair of x_1 and x_2 values are linked (as they are part of the same observation). Consequently, we need both x_1 and x_2 to be on the same graph as our response variable y – meaning we need to be graphing in three dimensions.

In a three dimensional graph the position of our data points is determined by a coordinate system involving (x_1, x_2, and y). What was our regression line is now stretched through 3 dimensional space to become a regression plane (Figure 16.2).

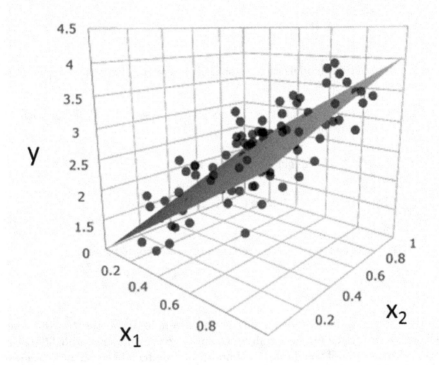

FIGURE 16.2

A regresssion plane

This plane is flat when our model is `lm(y~x1+x2)`, but becomes warped when we add an interaction like `lm(y~x1*x2)` (Figure 16.3). In the same figure we can see that the effect of an interaction model with a shared intercept `lm(y~x1:x2)` is to anchor the two sides of the regresssion plane where the two x-axes meet y-axis (at the value given by the intercept).

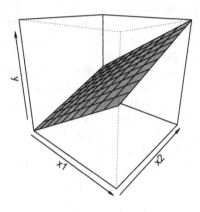

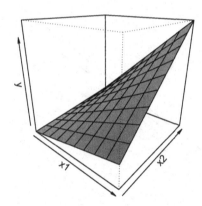

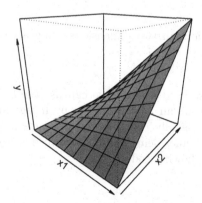

FIGURE 16.3
Regression planes when the two explanatory variables are both continuous

16.4.2.2 A visualisation trick

While it is hard to visualise interactions of continuous variables in multiple dimensions, you will recall how easy it was to visualise interactions involving categories. Consequently, when dealing with interactions, it is helpful to 'trick' R into handling a continuous variable like it was a category in order to visualise the effects of more complicated models.

We can do this by making a simplified data set in which each continuous variable is limited to a few values across its range. We can then use the *group* argument within a ggplot to treat these new values as categories. The result is a graph where we can predict y in terms of x_1 and x_2 without having to use more than 2 dimensions (Figure 16.4). In Figure 16.4 each of the categories corresponds to one of the horizontal lines present in the `lm(y~x1*x2)` regression plane (Figure 16.3). What we see is that while y increases with x_1 the value of y increases more at higher values of x_2.

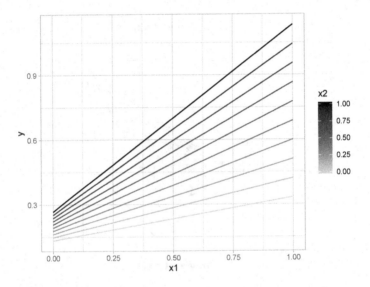

FIGURE 16.4
Visualising interactions in two dimensions using a continuous variable as a category

In order to make such a graph we can use the following steps below as a basic template. You can trial this code by using simple substitution. For example, we could visually check for an interaction (or lack of it – which would be signified by lines being parallel) between two x variables in the **hunt** data set using the following code:

```
x1 <- hunt$distance
x2 <- hunt$hours
y <- hunt$catch.weight
df <- data.frame(y, x1, x2)
```

Step 1: Simplify the data columns into discrete values using the **seq()** function in combination with sensible values across the variable's range. In this example, we use the **min()** and **max()** functions which would be appropriate in most situations. Importantly, we need to set a manageable number of divisions (using the *length.out* argument) as these will be the number of lines plotted in our graph. Because we are going to apply our original model to this new data we need to ensure of variables have the same names as those in our model formula:

```
x1 <- seq(from = min(x1), to = max(x1), length.out = 10)
x2 <- seq(from = min(x2), to = max(x2), length.out = 10)
```

Step 2: Create all possible combinations of the variables using the **expand.grid()** function. This will result in a data frame:

```
sim.data <- expand.grid(x1, x2)
```

Step 3: The variable names will be lost so we will have to rename them and save as a data frame:

```
colnames(sim.data) <- c("x1", "x2")
sim.data <- as.data.frame(sim.data)
```

Step 4: Now we need to predict the response variable from our model:

```
my.model <- lm(y ~ x1 * x2, data = df)
sim.data$y <- predict(my.model, newdata = sim.data)
```

Step 5: Make a ggplot using **geom_line()** where the group and colour arguments refer to a different explanatory variable than that given by the x argument in the aesthetics:

```
ggplot()+
  geom_line(data = sim.data, aes(x = x1, y = y,
                                 group = x2,
                                 colour = x2))
```

16.4.3 Colinearity and multicolinearity

It may seem obvious, but when we have a model with many explanatory variables we need to check that we are not accidentally measuring the same variable twice. For example, if we had two variables one measuring weight in grams and another measuring weight in kilograms these are the same measurement and our model should only include one of these. Similarly, we should avoid using explanatory variables which are highly correlated with each other (a situation known as colinearity) or can be predicted from combinations of other explanatory variables (known as multicolinearity). This is important because colinearity and multicolinearity can result in unreliable coefficient estimates.

An initial check for colinearity is to build a linear model with all our continuous variables and visually check for any linear relationships using the **pairs()** function. The **pairs()** function makes a grid of graphs allowing pairwise comparisons between continuous variables. In the example below we can check for issues in the **hunt** data set:

```
pairs(~age + distance + hours, data = hunt)
```

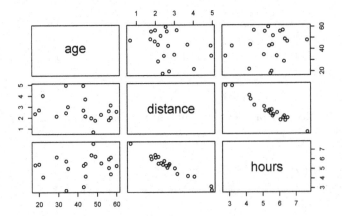

We can clearly see that the `distance` (from the hunting grounds) and `hours` (spent hunting) are correlated and appear to predict one another. However, a better way to check for colinearity/multicolinearity is to calculate a metric called the variance inflation factor (vif). A vif value greater than 5 signals a strong warning of colinearity/multicolinearity . We can use the **vif()** function from the `car` package to check:

```
library("car")
model.check <- lm(catch.weight ~ age + distance + hours,
                  data = hunt)
vif(model.check)

     age distance      hours
1.348381 29.764930 28.358035
```

What we see is that vif identifies both `distance` and `hours` contributing to colinearity/multicolinearity. In response we should remove one of these variables from our model (in this case `distance`), and recheck that the vif falls below 5:

```
model.check2 <- lm(catch.weight ~ age + hours, data = hunt)
vif(model.check2)

     age    hours
1.040007 1.040007
```

Clearly, the removal of `distance` has fixed the multicolinearity issues. The removal of a explanatory variable doesn't mean that it cannot be used in a model – just it shouldn't be used in the same model with the variables that it is colinear or multicolinear with.

16.5   Most statistical tests are linear models

Many people may find it surprising, but most common statistical tests are linear models. Older readers, like myself, were taught a large number statistical tests with vastly different names. As a consequence, many of us were unnecessarily confused – not realising that almost all of these tests were linear models which only differed in model formula.

R contains a large number of functions for these specific statistical tests. In the sub-sections below I show that these functions can be understood and closely replicated using the **lm()** function (as judged by the tests returning the same p value associated with the parameter of interest). The purpose of presenting the **lm()** equivalent is primarily to allow you to understand the test in terms of a model formula. It is worth noting that sometimes the specialty functions in R have a variety of arguments that cannot be easily replicated in the **lm()** function. However, given that most studies in conservation and development are observational rather than experimental the **lm()** function and its extensions will usually be adequate. However, if a truly experimental approach is being undertaken then one of R's specialty functions might be required[1].

In the past null hypothesis testing was the approach most people were taught. One of the limitations of this approach was that only two models were considered: the null model (in which the parameter of interest is left out) and the alternative model (in which the parameter of interest is included). Nowadays, thanks to the development of model selection approaches (demonstrated in Chapter 15) we can actively test a wide range of alternative models effectively comparing the explanatory power of models that were once considered entirely different tests. In addition, model selection allows us to compare models which are non-nested (meaning they don't have to use a subset of the same explanatory variables).

16.5.0.1 One sample t-test

One of the simplest tests is called a one sample t-test. This test is used to decide if the mean of our sample is different from an expected value (which is usually assigned the name *mu*). For example, we might be interested in testing whether or not the mean of our sample is consistent with a published value (which would be our *mu*). This test uses the **t.test()** function (note: in the one sample version, the argument *data* cannot be used within the **t.test()** function). In the linear model version of the test we subtract the expected value (*mu*) from our response variable:

```
lm(mu - y ~ 1)              t.test(your.data$y, mu = mu)
```

In the example below we are checking to see whether the mean of our sample variable, y, is consistent with a mean value of 10 (*mu*). The difference between the two means is given by the intercept in the linear model summary:

---

[1]in which a specific hypothesis is being tested in a highly controlled environment

```
lm(10 - y ~ 1, data = demo1) %>% summary()
```

Which is equivalent to:

```
t.test(demo1$y, mu = 10)
```

16.5.0.2 Independent t-test

An independent t-test is checks whether there is support for two groups having the same or different means (i.e. same or different intercepts) when there is just one categorical explanatory variable:

```
lm(y ~ group)                    t.test(y ~ group,
                                        var.equal = TRUE)
```

```
lm(y ~ group2, data = demo1) %>% summary()
```

Which is equivalent to:

```
t.test(y ~ group2, var.equal = TRUE, data = demo1)
```

Note: Under null hypothesis testing two groups which have means with non-overlapping confidence intervals will be significantly different (at the 0.05 threshold). However, overlapping confidence intervals themselves do not automatically indicate significance or non-significance.

16.5.0.3 One-way ANOVA

A one-way ANOVA (Analysis of Variance) tests whether there is support for differences between three or more groups when there is just one categorical explanatory variable. A one-way ANOVA with a category containing only 2 groups will produce the same results as an independent t-test. The standard test in R uses the **aov()** function. A comparison between the **lm()** and **aov()** functions can be made by comparing their ANOVA tables (which contains the p values associated with the parameters of interest) using the **Anova()** function from the `car` package (using the argument *type = 3*)[2]:

```
lm(value ~ group1)                aov(value ~ group1)
```

```
lm(y ~ group1, data = demo1) %>% Anova(type = 3)
```

Which is equivalent to:

```
aov(y ~ group1, data = demo1) %>% Anova(type = 3)
```

16.5.0.4 Two-way ANOVA

A two-way ANOVA tests differences between two categorical explanatory variables. It usually also examines whether an interaction between the explanatory variables could affect the response variable. The standard test in R uses the **aov()** function.

```
lm(value ~                        aov(value ~
    group1 * group2)                  group1 * group2)
```

```
lm(y ~ group1 * group2, data = demo1) %>% Anova(type = 3)
```

Which is equivalent to:

---

[2]the *type* argument is required is because there are 3 different types of method to calculate the sums of squares for ANOVA and the default method differs between **lm()** and **aov()** functions

```
aov(y ~ group1 * group2, data = demo1) %>% Anova(type = 3)
```

16.5.0.5   ANCOVA

An ANCOVA (an Analysis of Covariance) is similar to a two-way ANOVA but with one continuous explanatory variable and one categorical variable. An interaction term is usually included.

```
lm(y ~ x * group)                    aov(value ~
                                              x * group)
```

```
lm(y ~ x * group1, data = demo1) %>% Anova(type = 3)
```

Which is equivalent to:

```
aov(y ~ x * group1, data = demo1) %>% Anova(type = 3)
```

16.6   Generalised linear models

Generalised linear models (GLM) are an extension of linear models which allow for response variables which have distributions other than the normal distribution. GLMs make this possible through a link function. The link function is a transformation which allows the response variable to be sensibly predicted from the explanatory variables. The three most common distribution families are the normal (Gaussian) distribution, the binomial distribution, and the Poisson distribution. Standard linear models use a methodology known as ordinary least squares (OLS) to estimate model parameters. OLS works by minimizing the sum of squared residuals of the observed data. By comparison, GLMs use a methodology known as maximum likelihood estimation (MLE) which requires knowledge of the distribution type in order to solveestimate the optimal value of the parameters.

GLMs can be undertaken in R using the **glm()** function. This function uses the same formula notation as the **lm()** function but at a minimum requires the distribution family to be named with the *family* argument:

```
glm(y ~ x, family = "binomial")
glm(y ~ x, family = "poisson")
glm(y ~ x, family = "gaussian")
```

16.6.0.1   The importance of link functions

If we were to make a model using the **lm()** function for binary data (e.g. true-false, yes-no, alive-dead) we would end up with a nonsensical set of predictions which extend past the logical limits of the response variable (i.e. greater than or equal to 0, or less than or equal to 1 – see Figure 16.5). By using **glm()** with an appropriate link function (sensible defaults are automatically associated with the *family* argument) the model prediction and its confidence intervals never exceed the logical limit of the distributions (Figure 16.5).

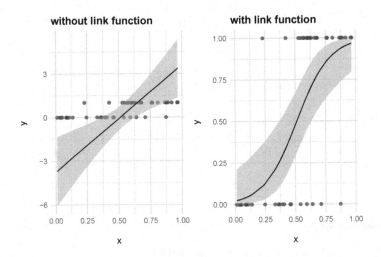

FIGURE 16.5

Left: incorrect model prediction using the incorrect link function for binomial data. Right: correct model prediction using the correct link function for binomial data. 95% confidence intervals shown in grey

In the following sections we shall examine the application of the **glm()** function to different distribution families. As this book is an introduction, we won't cover the different link function options available in **glm()** in any detail. However, as we are now exploring more complicated models, it is often useful to get summary information about the model – including the link function being used. The **model_info()** function of the insight package provides this, and other model information as a list (output not shown due to length):

```
library("insight")
glm(cases~intervention, family="poisson", data=prosecutions) %>%
  model_info()
```

16.6.1 Gaussian distribution (normal distribution)

As the Gaussian distribution is just another name for the normal distribution when we use the **glm()** function with the *family = "gaussian"* argument we will get exactly the same estimates as in **lm()**, despite the MLE method being used in **glm()** rather than the OLS method as in **lm()**:

```
lm(y ~ x + group1, data = demo1) %>%
  summary()
```

Which is equivalent to:

```
glm(y ~ x + group1, data = demo1, family = "gaussian") %>%
  summary()
```

16.6.2 Binomial distribution for logistic regression

If you recall the binomial distribution is used for a response variable with only two possible outcomes e.g. alive–dead, true–false, yes–no. When a binomial family is used in a GLM this is called a logistic regression. Logistic regression predicts the probability of a positive result (i.e. a '1'). As the predictions are constrained between 0 and 1, the line of best fit will have a tell-tale 'S' shape if the explanatory variable is continuous and it predicts the upper and lower limits of the response variable (see Figure 16.5).

We can demonstrate logistic regression with the **traps** data set. In this example we want to model the influence of **bait.type** (2 types: fish or egg) and **cover** (percentage forest cover) on **trap.catch** (whether or not a trap would be successful in catching a mongoose – a type of invasive predator):

```
logist.reg <- glm(trap.catch ~ bait.type + cover,
                  family = "binomial", data = traps)
```

Which returns the following coefficients:

```
coef(logist.reg)

      (Intercept)  bait.typeFish (dry)                  cover
    -0.0407399956       -0.4269926820          0.0002884887
```

The default link function for logistic regression is the logit function. The logit function calculates the probability (P) of the response variable being a '1' (i.e. successful). The linear equation for a logistic regression is:

$$\text{logit}(P) = \beta_0 + \beta_1 x_1 + \beta_2 x_2$$

Which is equivalent to the probability of the response variable being:

$$\log \frac{P}{1 - P} = \beta_0 + \beta_1 x_1 + \beta_2 x_2$$

Which if we convert back into the normal scale we are familiar with is:

$$P = \frac{e^{\beta_0 + \beta_1 x_1 + \beta_2 x_2}}{1 + e^{\beta_0 + \beta_1 x_1 + \beta_2 x_2}}$$

Which is the same as:

$$P = \frac{1}{1 + e^{-(\beta_0 + \beta_1 x_1 + \beta_2 x_2)}}$$

In the case of our top model the predictive equation would look like:

$$P(trap.catch) = \frac{1}{1 + e^{-(\beta_0 + \beta_1 \, bait.type + \beta_2 \, cover)}}$$

We can manually make our prediction for the first observation of the **traps** data set based on the value of its explanatory variables, and the values of our model coefficients:

```
traps$bait.type[1]

[1] "Egg"
```

```
traps$cover[1]

[1] 25.37894
```

We then substitute the coefficients (rounded to 6 decimal places) from our model and make a prediction for the first observation based on its explanatory variables. Note: bait.type uses the dummy variable '0' because it is not 'Fish (dry)':

```
1/ (1 + exp(-(-0.040740 -0.426993 * 0  + 0.000288 * 25.37894)))

[1] 0.4916431
```

We can then check this against R's **fitted()** function[3]:

```
fitted(logist.reg)[1]

        1
0.4916462
```

Potentially, these these fitted values could be rounded to predict the class of the response variable (i.e. '0' or '1'). As result, logistic regression can be also considered a type of classification.

16.6.2.1 Confidence intervals with link functions

As we saw in Figure 16.5, applying the correct link function is important to ensure our model makes sensible predictions. While getting model predictions is easy using the **fitted()** or **predict()** functions getting graphable confidence intervals is a little more complicated. This is because base R, as yet, does not automatically back-transform confidence intervals to the scale we are familiar with. However, we can use an R package (such the ciTools package and its **add_ci()** function) to automatically estimate these. Alternatively, we can make them manually by back-transforming the confidence intervals using the inverse of the link function. We can do this in a series of steps:

Step 1: In order to estimate the confidence interval around our prediction we

---

[3]a slight difference is caused by our use of the summary coefficients which are rounded to 6 decimal places

need to use the **predict()** function. Importantly, we need to make sure we use the arguments *type = "link"* and *se.fit = TRUE*.

```
probs <- predict(logist.reg, type = "link", se.fit = TRUE)
```

Step 2: Calculate the fitted, upper, and lower confidence intervals, by multiplying the critical value (which corresponds to a 95% confidence interval) with the standard errors from the prediction:

```
critical.value <- 1.96
fit <- probs$fit
upr <- probs$fit + (critical.value * probs$se.fit)
lwr <- probs$fit - (critical.value * probs$se.fit)
```

Step 3: As these predicted values are on the logit scale we need to back-transform them (into the normal scale that we are familiar with) using information stored within our model via `$family$linkinv()`. For our model object `logist.reg` this is:

```
traps$fit <- logist.reg$family$linkinv(fit)
traps$upr <- logist.reg$family$linkinv(upr)
traps$lwr <- logist.reg$family$linkinv(lwr)
```

Step 4: Plot the fitted values and confidence intervals using the **geom_ribbon()** function in ggplot:

```
ggplot()+
    geom_line(data = traps, aes(x = cover, y = fit,
                              linetype = bait.type))+
    geom_ribbon(data = traps, aes(x = cover, ymin = lwr,
                              ymax = upr,
                              group = bait.type),
                fill = "grey", alpha = 0.5)
```

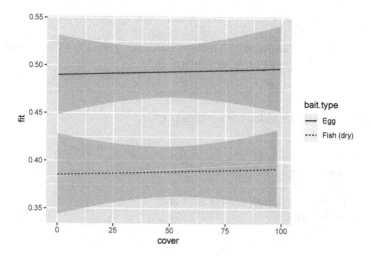

In the graph above we can clearly see that the egg bait type appears to be more effective than the fish bait type, and that forest cover has only a minor influence, if any, on the catch probability ('fit'). Because the continuous variable (`cover`) in this example is not a strong predictor of the response variable an 'S'-shaped prediction line is not produced.

16.6.3 Poisson regresssion for count data

When we are working with count data we are working with the Poisson distribution. Compared to binomial data, patterns in count data used for Poisson regression are usually easier to visualise. This is because the raw counts exist on the normal number scale (as opposed to being 0 or 1). There are a couple of important points to know:

- The count data used in a Poisson regression can only be zero or a whole number (never a decimal).

- If you want to model a density or rate by dividing count data by a measure of effort (e.g. time, or area) don't. Rather, leave the raw count data as the response variable but handle the scaling through the *offset* argument by placing it within the **log()** function. For example, if we were interested in differences between groups but our count data involved time periods which varied between observations the form of our model formula would be:

```
glm(count ~ group, offset = log(time), family = "poisson")
```

In the example below (using the `prosecution` data set) we are wanting to know whether more environment cases went before the village magistrate for prosecution after our intervention. As the variable `cases` is a count – the analysis requires a Poisson regression:

```
poisson.reg <- glm(cases ~ intervention, family = "poisson",
                   data = prosecutions)
```

Which returns the following coefficients:

```
coef(poisson.reg)

    (Intercept) interventionAfter
    -0.05129329        0.75273928
```

The default link function for Poisson regression is the log function[4]. The linear equation for a Poisson regression with two explanatory variables is:

$$\log(y) = \beta_0 + \beta_1 x_1 + \beta_2 x_2$$

Which is equivalent to:

$$y = e^{\beta_0 + \beta_1 x_1 + \beta_2 x_2}$$

In the case of our example, we have a model with only one explanatory variable, therefore:

$$cases = e^{\beta_0 + \beta_1 intervention}$$

We then substitute the coefficients (rounded to 6 decimal places) from our model and make a prediction for the first observation based on the type of intervention. We use the dummy variable '0' because the intervention is not 'After':

```
exp(-0.05129329 + (0.75273928 * 0))  # intervention = Before

[1] 0.95
```

And check this against R's **fitted()** function:

---

[4]for this reason they are also known as log-linear models

```
fitted(poisson.reg)[1]

    1
0.95
```

16.6.3.1 Diagnostics for GLM

While the standard residual plots given by R are good for linear models and Gaussian models they are much harder to interpret for other GLMs. Often people will mistakenly think these plots suggest there are major problems when there are none. The DHARMa package provides diagnostic plots which are easier to interpret for GLMs. These diagnostic plots are based around standardised residuals[5] made from simulated data. Consequently, our first step is to simulate the residuals of our model using the **simulateResiduals()** function:

```
library(DHARMa)
poisson.residuals <- simulateResiduals(poisson.reg)
```

We can then plot these simulated residuals which results in two plots: a 'QQ plot residuals' which can be interpreted in the same way as the 'normal Q-Q plot' from the standard diagnostic plots (see Chapter 15), and a plot of the 'Residual vs predicted'. The 'QQ plot residuals' also provides the outcome of a Kolmogorov-Smirnov test ('KS test') which tests whether the residuals are uniform across the predicted values. The same plot also shows a significance test for outliers (which will indicate if some of our observations were larger or smaller that expected), and whether there is over- or under-dispersion (see section 16.6.3.2).

Ideally, in the 'Residual vs predicted' graph the boundaries and mid-point of the box plot for the categorical explanatory variables should align with the dotted lines at 0.25, 0.5, and 0.75 on the y-axis:

```
plot(poisson.residuals)
```

---

[5]standardised residuals are the residuals divided by the standard deviation of the residuals

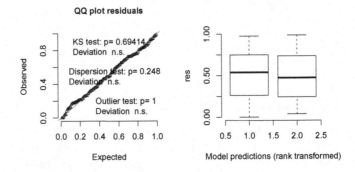

A model with either only continuous explanatory variables or a mixture of variable types will result in a scatterplot rather than a box plot. In this scatterplot red trend lines[6] should ideally align with the 0.25, 0.5, and 0.75 guidelines on the y-axis (Figure 16.6). Possible outliers will also be marked by an asterisk. Importantly, if we are working with a small sample close alignment with the y-axis guide lines is unlikely, and the position of red trend lines should be interpreted as a rough guide only. For example, in Figure 16.6, the curved shape of the lower line is being picked up as a possible issue (and will appear coloured red). If we are concerned about the outcome of an outlier test, or the position of any of the red lines we should first increase the number of samples used by the **simulateResiduals()** function by setting the n argument higher than the 250 default (e.g. $n = 1000$) and then check if the problem persists.

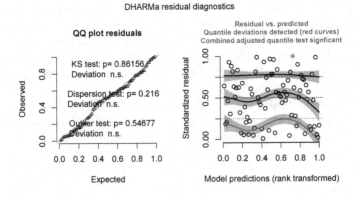

FIGURE 16.6
DHARMa plot showing deviation of residuals

---

[6]a result of a quantile regression (a method not covered in this book)

16.6.3.2 Under and over-dispersion

The Poisson distribution is based on a single parameter which is both the mean and variance, as a consequence the mean should equal the variance. This is the key assumption of Poisson regression. When the variance is greater than the mean of our data, then it is called over-dispersion, if it is less, it is called under-dispersion. Over-dispersion is much more common than under-dispersion. A simple test of under or over-dispersion is through the **testDispersion()** in the DHARMa package which also makes a diagnostic plot. A low p value from this test would suggest that our model is not meeting this key assumption, as would a plot where the vertical red line appears close to, or beyond the margins of the distribution (which is not the case in this model):

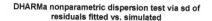

```
testDispersion(poisson.residuals)

        DHARMa nonparametric dispersion test via sd of
        residuals fitted vs. simulated

data:  simulationOutput
ratioObsSim = 0.92059, p-value = 0.248
alternative hypothesis: two.sided
```

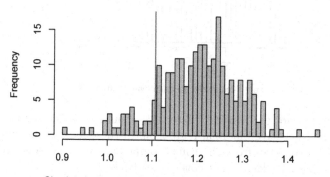

Simulated values, red line = fitted model. p−value (two.sided) = 0.248

Should such tests result in issues an alternative modelling route is to use a negative binomial regression. A negative binomial regression adjusts the variance independently from the mean via an extra model parameter (e.g. **glm.nb()** from the MASS package – not covered in this book).

Another issue common to count data is an over abundance of zero counts. This is a fairly common situation in wildlife monitoring. This can be tested in the DHARMa package through the **testZeroInflation()** which undertakes a test and gives a diagnostic plot. A low p value from this test would suggest that our model is zero-inflated, and the resulting plot would have a vertical red line appearing close to, or beyond the margins of the distribution (which is not the case in this model):

```
testZeroInflation(poisson.residuals)

        DHARMa zero-inflation test via comparison to
        expected zeros with simulation under HO = fitted
        model

data:  simulationOutput
ratioObsSim = 0.83088, p-value = 0.296
alternative hypothesis: two.sided
```

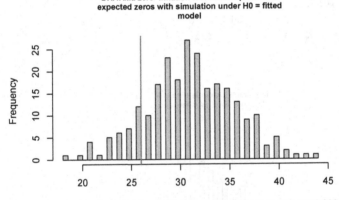

Simulated values, red line = fitted model. p-value (two.sided) = 0.296

If our model appears to be zero-inflated we should consider modelling using a zero-inflated negative binomial regression. In these models the excess zeros are modeled separately from the count component (e.g. the **zeroinfl()** function from the pscl package – not covered in this book).

16.6.4   Chi-squared tests

Often when we have count data we want to know if there are differences between groups in terms of observed and expected frequencies. For example, in conservation and development projects we are often concerned about differences in participation rates between different groups of people (e.g. men and women). Chi-squared tests can tell us if our observed rates differ from those we expected. For example, if we ran some training in which 131 women and 203 men attended we could test whether this result was consistent with an equal sex ratio (i.e. no differences between the proportions of two groups attending training). We can test this with a goodness-of-fit test (see section 16.6.4.2 for more detail):

```
chisq.test(c(131, 203))

        Chi-squared test for given probabilities

data:  c(131, 203)
X-squared = 15.521, df = 1, p-value = 8.16e-05
```

Which results in a p value so close to zero we can be very confident that the sexes did not have the same probability of attending (as the p value is so low it would be highly unlikely to achieve this result by chance). However, if we wanted to, we could check to see if this pattern was consistent with a different ratio (e.g. a female:male ratio of 40:60) by giving a set of expected proportions to the argument p:

```
chisq.test(c(131, 203), p = c(0.4, 0.6))

        Chi-squared test for given probabilities

data:  c(131, 203)
X-squared = 0.084331, df = 1, p-value = 0.7715
```

Which results in a p value suggesting that such a result (or one more extreme) would be obtained ~77% of the time. Hence we conclude that our counts are consistent with a female:male ratio of 40:60.

16.6.4.1 Data preparation

When we are working with participant lists and want to prepare our data for a chi-squared test there are two possible data formats:

1. A table which shows the raw counts separated by category (this is known as a contingency table). Contingency tables can be created directly from the data via the **table()** function (shown here using the **attendees** data):

```
table(attendees$gender)

female    male
   131    203
```

We could also separate gender by the type of course (thereby testing whether the two categorical variables behave independently – this is known as a test of independence see section 16.6.4.3 for more detail). The table can then be passed to the **chisq.test()** function:

```
chisq.test(table(attendees$gender, attendees$course))

        Pearson's Chi-squared test

data:   table(attendees$gender, attendees$course)
X-squared = 1.0025, df = 2, p-value = 0.6058
```

Or it could be created manually using the **c()** function in the case of a contingency table with one variable, or using the **matrix()** function in the case of a contingency table with two variables:

```
chisq.test(c(131, 203))
chisq.test(matrix(c(36, 52, 43, 56, 90, 57), ncol = 2))
```

2. Alternatively, if two variables are being used these can be directly passed to the **chisq.test()** function – this however, will not work with only one variable (note: unlike most other modelling functions **chisq.test()** does not use a *data* argument):

```
chisq.test(attendees$gender, attendees$course)
```

As already mentioned there are two common types of chi-squared test. Each of these tests can also be expressed as a comparison between two Poisson regressions. While it is far easier to use the built in chi-squared functions it is helpful to understand the comparison we are making we in terms of a GLM.

16.6.4.2 Chi-squared goodness-of-fit test

The Chi-squared test goodness-of-fit is similar to a one-way ANOVA but with count data. In this model we are comparing two models with null hypothesis testing: an intercept only model (H0: our null hypothesis), and a model which assumes there are differences between the frequency of categories (Ha: our alternative hypothesis):

```
H0 <- glm(count ~ 1,              chisq.test(df$count)
          family = "poisson")
Ha <- glm(count ~ group,
          family = "poisson")
```

In order to test this equivalence we will need to produce a summary of the count data for the **glm()**:

```
one.variable <- attendees %>%
  group_by(gender) %>%
  summarise(count = n())
```

We can then show this equivalence using the **anova()** function using the argument *test = "Rao"*:

```
H0 <- glm(count ~ 1, family = "poisson",
          data = one.variable)
Ha <- glm(count ~ gender, family = "poisson",
          data = one.variable)

anova(H0, Ha, test = "Rao")
```

Which matches the chi-squared test:

```
chisq.test(table(attendees$gender))
```

16.6.4.3 Test of independence

The chi-squared test of independence is like a two-way ANOVA but with count data. In this test we compare two models: an additive model (H0: our null hypothesis), and an interaction model (Ha: our alternative hypothesis) in which the variables interact – and are therefore not independent of each other.

```
HO <- glm(count ~                        chisq.test(df$count)
          group1 + group2,
          family = "poisson")
Ha <- glm(count ~
          group1 * group2,
          family = "poisson")
```

In order to test this equivalence we will need to produce a summary of the count data for the **glm()**:

```
two.variables <- attendees %>%
  group_by(course, gender) %>%
  summarise(count = n())
```

We can then show this equivalence using the **anova()** function using the argument *test = "Rao"*:

```
HO <- glm(count ~ gender + course, family = "poisson",
          data = two.variables)
Ha <- glm(count ~ gender * course, family = "poisson",
          data = two.variables)

anova(HO, Ha, test = "Rao")
```

Which matches the chi-squared test:

```
chisq.test(attendees$gender, attendees$course)
```

Disaggregation
Donor reports often ask for data to be 'disaggregated' by gender or some other categorical variable. 'Disaggregated' simply means separating the total count data by a specific category. The **table()** function is the easiest way to do this for one or two variables. Chi-squared tests then provide a robust way of assessing whether such separation is statistically justified. However,

chi-squared testing and the **table()** function are only useful for two or fewer categorical variables.

For disaggregation with more than two categorical variables I would recommend using the **group_by()** and **summarise()** functions from the tidyverse package to produce tables, and using Poisson regression to understand differences between models.

Often when people disggregate training data they treat it as though the people attending the different courses are always different. In some situations this would be true (e.g. courses run in different cities) but for projects working in a village setting often this is not the case. Often training is attended by the same core group of people and this could lead to over counting. For example, a person attending 5 different courses is not the same as 5 different people all attending 1 course. Consequently, when disaggregating, it is often more appropriate to first summarise the data by the unique individuals attending (e.g. by name or some other identifier). We can demonstrate the difference below:

```
attendees %>%
  group_by(village) %>%
  summarise(count = n())

# A tibble: 5 x 2
  village     count
  <chr>       <int>
1 Anota          93
2 Lamaris        34
3 Maniwavie      41
4 Nugi           87
5 Takendu        79

attendees %>%
  group_by(village) %>%
  summarise(count = n_distinct(name))

# A tibble: 5 x 2
  village     count
  <chr>       <int>
1 Anota          46
2 Lamaris        17
3 Maniwavie      18
4 Nugi           40
5 Takendu        35
```

16.7 Other related modelling approaches

There are a number of other related modelling approaches which are useful for common conservation and development projects. In this section we look briefly at repeated measures (useful when we are repeatedly measuring the same replicate through time), cumulative link models (useful in analysing Likert data which is acquired in questionnaires), beta regression (useful for analysing true proportions), and non-parametric equivalents to common classical statistical tests (useful if assumptions of linear models are violated).

16.7.1 Repeated measures

A lot of modelling (and experimentation) is based on the concept of replication. Replication ensures we have enough samples (replicates) in a treatment group so that we can identify the effect of our treatment (i.e. our interventions) from the natural variability of our the samples. Ideally, we would want the replicates to be exactly the same – this way we could remove all variability.

While scientists working in laboratory conditions can make almost exact replicates, it is almost impossible to have exact replicates in non-laboratory situations, and as a consequence there is nearly always variability between our replicates (this why medical studies using twins are so popular).

Up until now we have handled the issue of variability by ensuring we choose our samples randomly and have a large enough sample size. However, a different way of handling this problem is by ensuring we use the same replicates and monitor their response to a treatment through time. This is known as a repeated measures design. In a repeated analysis we want to untangle the random effects (variability) associated with our replicates from the effects we are really interested in (called fixed effects). For this reason, when we are making models using repeated measures we need to describe the structure of the replicates.

For example, if were studying the long-term effectiveness of different teaching methods we might repeatedly measure the exam performance of the same school children across a number of schools. Before we even start the study we know differences will exist between individual children (because of different learning abilities) and different schools (because of past performance). In this case, both the variables of child and school would be considered random effects. As these variables are not what we are really interested in, they are sometimes called nuisance variables.

There are two popular packages which can handle this kind of data in R: `nlme` and `lme4`[7]. In the examples below I use the `nlme` package because it clearly separates out the random effects, and reports *p* values.

---

[7] `lme4` has advantages over `nlme` including the ability use different distributions via the **glmer()** function and integration with the `DHARMa` package

In the **lme()** function[8] the fixed effects are separated from the random effects. The fixed effects come first and appear as a standard model formula (in the example below: `score~training`). In the random effect formula the structure of the replicate appears after the '|' sign. For example a child from a particular school would be:

```
lme(score ~ training.method, random = ~1|school/child)
```

In this situation we would say that `child` is nested within `school`. The hierarchy of the replicate system is separated using the '/' sign (in descending order from the left). If all our samples of children were from one school, we would not need the school variable, and the formula would be:

```
lme(score ~ training.method, random = ~1|child)
```

The example above with the code **random=** `~1|child` represents an intercept only model. The intercept model is just like the null model from a normal model formula (y~1) except it occurs before the '|' sign. We can think of this as each child starting from a different position in terms of their exam performance. As an intercept only model, we are looking for an effect associated with the training method even though the children could be of different abilities at the start.

Alternatively, we might expect that each child might react differently to the training method depending on their age at the time. We can think of this as a normal linear model formula (with an intercept and slope: y ~ **age**). This model formula is then placed before the '|' sign. Now we have a model which assesses the effect of the training method knowing that each child will begin from a different starting point (intercept) and possibly respond differently according to their age (slope):

```
lme(score ~ training.method, random = ~age|child)
```

If we are interested in model selection for the fixed effects we need to use the argument *method = "ML"* and keep the random effects constant in our candidate models (assuming we have created a list object called **mod** for our models – see Chapter 15). This might look like:

```
mod[[1]] <- lme(score ~ 1, random = ~age|child,
                method = "ML")
mod[[2]] <- lme(score ~ training.method, random = ~age|child,
                method = "ML")
```

However, if we are interested in model selection for the best explanation of random effects we need to keep the fixed effects constant and use the argument *method = "REML"*.

---

[8]short for linear mixed effects (i.e. the mix of fixed and random effects)

```
mod[[1]] <- lme(score ~ training.method, random = ~1|child,
            method = "REML")
mod[[2]] <- lme(score ~ training.method, random = ~age|child,
            method = "REML")
```

16.7.1.1   Pairwise t-test

The simplest type of repeated measures analysis is known as a pairwise t-test. In this test the sample subjects are measured twice (i.e. the group would normally be 'before' or 'after'):

```
lme(y ~ group,                    t.test(y ~ group,
     random = ~1|subject,                paired = TRUE,
     data = df)                          data = df)
```

To demonstrate this example we will make a reproducible example. In this example there are twenty subjects (which each have a personal intercept). Each subject is measured before and after the project intervention:

```
set.seed(19)                          # random seed
subject <- rep(letters[1:20], 2)      # subject name
subj.int <- rep(rnorm(20,0,1), 2)     # subject intercept
period.dum <- c(rep(0, 20), rep(1, 20)) # period (dummy var)
period <- c(rep("Before", 20),        # period (name)
            rep("After", 20))
intercept <- 10                       # intercept
slope <- 2                            # slope
error <- rnorm(40, 0, 2)              # error

# make  response variable
y <- intercept + slope * period.dum + subj.int + error

# make data frame
pair.df <- data.frame(subject, period, slope, y)
```

The output of **lme()** summary:

```
lme(y ~ period,  random = ~1|subject, data = pair.df) %>%
    summary()
```

Closely approximates:

```
t.test(y ~ period, data = pair.df, paired = TRUE)
```

16.7.2 Cumulative link models

Frequently, in conservation and development projects we collect ordinal (ordered) categorical data. This kind of data is made up of categories that have a clear order. Often a 5-point scale (known as a Likert scale) is used to record this kind of information. Ordinal categories are often used with regard to agreement (e.g. strongly disagree–strongly agree), satisfaction (e.g. very dissatisfied–very satisfied), frequency (e.g. never–always), or concepts of quantity (e.g. a little–a lot). Ordinal data is also very useful for gathering information from communities with low numeracy.

The analysis of ordinal data can be done through an extension to linear modelling called cumulative link models (also known as ordinal regression). The most important step is to code the response variable as an ordered factor. We can do this through the **factor()** function, giving the order of the categories through the *levels* argument (if not already ordered) and using the *ordered = TRUE* argument:

```
y <- factor(y, ordered = TRUE,
            levels = c("never", "occasionally",
                       "sometimes", "often", "always"))
```

Let's begin by simulating some Likert data. It should be pointed Likert data is just a way of gathering data – it is not a distribution. Therefore we can simulate the data using what ever distribution we think best matches the underlying process. In the following example we'll use a linear model to simulate the data.

To begin we'll make a response variable (**response**) using a linear equation with **female** as a categorical dummy variable (female ='1', male = '0') and **age** as a continuous variable. We then convert the response variable into a Likert scale by rounding it and changing it to an ordered factor:

```
set.seed(10)
# make variables
n = 100
age <- round(rnorm(n, 40, 9),0)
female <- rbinom(n, size = 1, prob = 0.5)
```

```
# linear equation
response <- 0 + 0.4 * female + 0.07 * age + rnorm(n, 0, 1)
# round and change response to factor
response <- as.factor(round(response, 0))
# make data frame
df <- data.frame(response, female, age)


# rename factors
df <- df %>% mutate(response = fct_recode(response,
                                    "v.poor" = "1",
                                    "poor"   = "2",
                                    "ok"   = "3",
                                    "good"  = "4",
                                    "v.good"  = "5"))
# make ordered factor
df$response <- factor(df$response, ordered = TRUE)
# change female to factor
df$female <- as.factor(df$female)
```

We then analyse the data using the clm() function from the ordinal package. Understanding the coefficients from a cumulative link model is somewhat harder than from a GLM. The cumulative link model not only produces coefficients (βs) – which we are familiar with, but also threshold coefficients (αs):

```
library(ordinal)
our.model <- clm(response~ age + female, data = df)
our.model

formula: response ~ age + female
data:     df

 link  threshold nobs logLik  AIC     niter max.grad cond.H
 logit flexible  100  -139.74 291.47 5(0)  8.65e-09 2.5e+05

Coefficients:
    age female1
 0.1102  0.8929

Threshold coefficients:
```

v.poor|poor	poor|ok	ok|good	good|v.good
2.440	4.324	5.739	7.529

These coefficients are used to calculate the cumulative probability of being beneath an ordinal class (e.g. poor) given by the equation. You will notice that this equation appears similar to the equation used in logistic regression (see section 16.6.2). This is because both methods use the logit link function. However, this equation differs in that there is subtraction occurring between the threshold coefficients and the other regression coefficients:

$$P(Y \leq poor) = \frac{1}{1 + e^{-(\alpha_{v.poor|poor} - \beta_1 age - \beta_2 female1)}}$$

As a result, we can manually calculate the cumulative probability of being beneath the v.poor|poor boundary (given by the threshold coefficient 2.440) for our first observation – a 40 year old man i.e. `age = 40`, `female1 = 0`. As we are using rounded coefficients, there will be a small difference between the cumulative probabilities we calculate and those given by the **predict()** function later:

```
1/(1 + exp(-(2.440 - 40 * 0.1102 -  0.8929 * 0)))

[1] 0.1226039
```

However, if we wanted to predict the probability of an observation falling in a particular category we need to calculate the cumulative probabilities for all thresholds. Rather than manually calculate the logit values we can use the **plogis()** function for the different threshold coefficients:

```
b1 <- plogis(2.440 - 40 * 0.1102 - 0.8929 * 0)
b2 <- plogis(4.324 - 40 * 0.1102 - 0.8929 * 0)
b3 <- plogis(5.739 - 40 * 0.1102 - 0.8929 * 0)
b4 <- plogis(7.529 - 40 * 0.1102 - 0.8929 * 0)
```

We then subtract the cumulative probabilities for the interval we are interested in. The results of the model suggest that the category of **response** for our observation will be in the category 'poor':

```
c(v.poor = b1 - 0,
  poor = b2 - b1,
  ok = b3 - b2,
  good = b4 - b3,
  v.good = 1 - b4)

   v.poor        poor          ok        good      v.good
0.12260387 0.35640847 0.31199366 0.16674471 0.04224929
```

We can also predict the category directly using the **predict()** function with *type = "class"* argument:

```
pred.categories <- predict(our.model, type = "class")
```

Which also predicts the category of the first observation to be 'poor':

```
pred.categories$fit[1]

[1] poor
Levels: v.poor poor ok good v.good
```

Rather than the using the manual method, we can quickly generate a table with the code below (there will be a slight difference from our manual results due to rounding):

```
example <- data.frame(age = 40, female = as.factor(0))
predict(our.model, example, type = "prob")$fit

    v.poor        poor          ok        good      v.good
1 0.1228052 0.3566691 0.3118009 0.1665572 0.04216765
```

16.7.3   Beta regression

In earlier sections, we covered working with binary data (logistic regression) and working with count data including densities and rates (Poisson regression). But what if we are working with true proportions made by dividing a continuous variable by another continuous variable e.g initial weight/final weight? Historically these types of proportions were handled by transforming the

TABLE 16.1

Classical non-parametric tests and their parametric approximations

Test name	R function	Alternative to	Equivalent to
Wilcoxon signed-rank	wilcox.test(y)	lm(y ~ 1)	One sample t-test
Mann-Whitney U	wilcox.test(y~group)	lm(y ~ group)	Independent t-test (2 groups only)
Kruskal-Wallis	kruskal.test(y~group)	lm(y ~ group)	One way ANOVA (3+ groups)

response variable[9] and applying a linear model. In the last couple of decades a new type of analysis, called beta regression, has been popularised which avoids some of the pitfalls associated with the historic approach. Beta regression can be undertaken in R using the **betareg()** function from the **betareg** package. However, as data based on proportions is rarely encountered in conservation and development projects (as most monitoring data is focused on counts) I do not cover beta regression here.

16.7.4 Non-parametric approaches

Thus far we have only looked at modelling approaches which involve families of distributions. However, there are a large number of tests which don't rely upon known probability distributions, nor the assumptions of linear models. These are known as non-parametric approaches. This does not mean they have no parameters but rather they are not constrained by a statistical distribution.

Many of the classic non-parametric tests mimic common linear models (see Table 16.1). These non-parametric methods analyse data based on their ordered ranking. These classic non-parametric tests are only used for null hypothesis testing – and cannot be used for model selection.

Let's imagine a group of villagers claim that women in their village were not disadvantaged by poorer educational opportunities because a woman had managed to get a university education. We doubt this conclusion so we randomly survey 12 adults: 6 females and 6 males chosen at random[10]. We are interested in the years of schooling each individual received:

```
years <- c(1, 2, 3, 4, 5, 6, 7, 8, 9, 10, 19, 20)
gender <- c("F","F","F","F","F","M","M","M","M","M","F","M")
schooling <- data.frame(years, gender)
```

---

[9]typically using an arcsine transformation

[10]this small sample size is solely for purposes of a manageable example

We can imagine that we first attempt to analyse the data using a linear model: lm(years~gender, data=schooling). However, when we check the Normal Q-Q plot in plot diagnostics we notice the residuals aren't following the normal distribution (due to two people undertaking advanced degrees). So we try the non-parametric Mann-Whitney U test which is the non-parametric version of our linear model using the **wilcox.test()**:

```
wilcox.test(years ~ gender, data = schooling)

        Wilcoxon rank sum test

data:  years by gender
W = 5, p-value = 0.04113
alternative hypothesis: true location shift is not equal to 0
```

The Mann-Whitney U test results in a p value of 0.04113, a rather low value to achieve by chance (lower than the 0.05 traditionally used as a threshold) consequently, we might conclude there is a difference between the schooling opportunities between females and males in this particular village. However, as there is a well known global pattern of women being excluded from educational opportunities we would have been justified in choosing the one-tailed version of the test (where the argument *alternative* = *"less"* is used). A one-tailed test only examines the probability that a specific sample group is either consistently larger or smaller than the other, as opposed to the 'two-tailed' default which doesn't assume a specific direction.

```
wilcox.test(years ~ gender, data = schooling,
            alternative = "less")

        Wilcoxon rank sum test

data:  years by gender
W = 5, p-value = 0.02056
alternative hypothesis: true location shift is less than 0
```

On this occasion our one-tailed test has resulted in a p value half the size of the original – thereby reducing the probability that this result (or one more extreme) could be achieved by chance if there was no difference between the two groups. Generally, one-tailed tests are to be avoided unless you have a very strong basis for the direction of the result prior to the test. By using a one-tailed test we are risking missing a difference existing in the opposite direction (i.e. females having more years of schooling than males).

16.7.5 Advantages of non-parametric tests

One of the big advantages of these non-parametric tests is that when we use them we don't need to worry about the assumptions of linear models. All of the classical tests mentioned in this section are based on ordered ranks – not the actual value of the response variable. If we look at the rank of our **years** variable using the **rank()** function we get the following output:

```
rank(years)

[1]  1  2  3  4  5  6  7  8  9 10 11 12
```

As the test is working on ordered ranks (not actual values) we can change the value of the variables and so long as the ranks do not change the result will stay the same. For example, we could change the second to last value of **years** to 13 and the last value of **years** from 20 to 237 and it still won't change the result. The downside of these classical non-parametric methods is that they focus on null hypothesis testing and can not be used for prediction (or model selection).

16.8 Recommended resources

- https://lindeloev.github.io/tests-as-linear/. A guide to expressing common statistical tests as linear models by Jonas Kristoffer Lindeløv.

- https://marissabarlaz.github.io/portfolio/ols/. A guide for understanding cumulative link models (ordinal logistic regression) by Marissa Barlaz.

- https://rcompanion.org/handbook/E_01.html/. A guide to analysing Likert data by Salvatore S. Mangiafico.

16.9 Summary

- Generalised Linear Models (GLMs) are an extension of linear modelling that allows for different kinds of distributions.

- Many well known statistical tests are just a specific type of linear model or an extension of the linear modelling approach.

- A few key characteristics of the response variable can help us recognise the most appropriate linear modelling approach (see Figure 16.7).

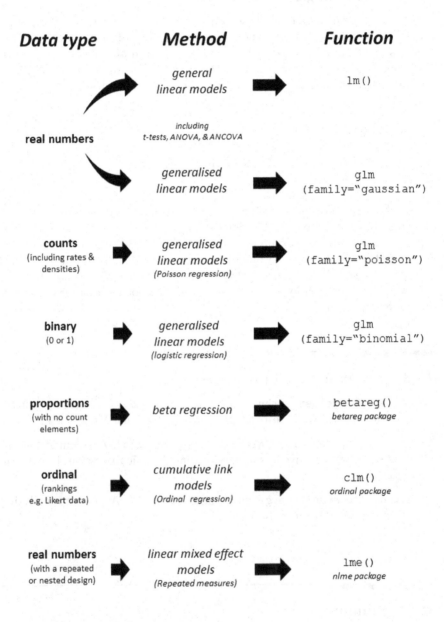

FIGURE 16.7
A basic guide to linear modelling approaches

17

Introduction to clustering and classification

In this chapter we will introduce:

- Supervised versus unsupervised learning.

- Clustering.

- Dimension reduction (a way of simplifying variables).

- Classification using classification trees.

17.1   Clustering and Classification

In the previous chapters we have focused on regression. Although regression is a powerful technique it can't solve certain types of questions. This is why some understanding of clustering and classification is necessary. Clustering refers to techniques which identify hidden groupings in data, while classification refers to techniques that can predict a categorical response variable. While these techniques have a wide variety of uses they are particularly useful for predicting and understanding patterns in questionnaire data. Clustering and classification are huge fields, and for this reason this chapter will only introduce a few basic concepts.

17.2   The packages

In this chapter we will require a number of packages. We will introduce clustering using base R functions along with the dimension reduction methods from `FactoMineR` with help from `factoextra` to graph its outputs. For the classification section of the chapter we will use the `rpart` package to introduce classification trees, and `rpart.plot` to plot the outputs. The `caret` package will be introduced as a platform for machine learning. We will also use the `tidyverse` package for some data wrangling.

```
library("caret")
library("factoextra")
library("FactoMineR")
library("rpart")
library("rpart.plot")
library("tidyverse")
```

Will also require the `randomForest` package to be installed, but not necessarily loaded, so that we can demonstrate how we can use the `caret` package to make model comparisons between different machine learning techniques.

```
install.packages("randomForest")
```

17.3   The data

For this chapter we will analyse a simplified example of questionnaire data using the `questionnaire` data set:

```
library("condev")
data(questionnaire)
```

17.4   Supervised versus unsupervised learning

Supervised learning refers to a situations where we are trying to predict a response variable from a set of explanatory variables. Linear and generalised linear models are therefore examples of supervised learning. When we try to predict a continuous response variable we call it regression, but when we are trying to predict a categorical response variable we call the process classification. By comparison unsupervised learning does not have a response variable. Instead unsupervised learning attempts to find some kind of grouping structure in the data. These groupings are known as clusters, and as a result the technique is often called clustering. To summarise:

- We use supervised learning to predict a response variable from a set of explanatory variables. There are two types of supervised learning:

  - Regression: when we are predicting a response variable which is a continuous (i.e. numeric).

  - Classification: when we are predicting a response variable which is a category (i.e. a factor)

- We use unsupervised learning to assign observations to groups.

Classification and clustering are large topics and cannot be covered in a single book, yet alone a single chapter. So in this chapter we will be only able to lightly touch on clustering and classification.

17.4.1 Why learn classification and clustering

Classification and clustering are used extensively by various disciplines and businesses to understand why people use certain products and how these products can be better marketed to them. In a similar way, in conservation and development, we are usually interested in understanding how our inventions can be better tailored to certain groups and become more effective.

A quick way of obtaining large amounts of information from a group of people is through a questionnaire[1]. We most commonly use questionnaires to collect information from communities and other project stakeholders at the beginning of a project to inform its design, or at the end to evaluate its performance. Sometimes we want to use this data to identify groups of people based on how they are answering the questions. We can then use this information to better target our future projects to these groups. In such a situation we would use clustering. Alternatively, we might want to understand what experiences cause people to answer in a particular way. Here, we would be trying to predict how people answer a particular question based on their other answers – consequently we would want to use a classification technique.

17.5 Clustering

Clustering is often used as a form of data exploration to help us understand hidden patterns or relationships within the data. A good clustering method

---

[1]for purposes of this chapter we are only considering answers from closed questions (those which only allow a response which fit into pre-arranged categories)

will make clusters which have lots of internal similarity but which share little similarity with other clusters. There are many different kinds of clustering methods. In this chapter we will introduce cluster analysis using hierarchical clustering. We will also introduce dimension reduction to show how this technique can be used to help analyse complex questionnaire data.

17.5.1   Hierarchical clustering

Hierarchical clustering works out the clusters in a step-wise process by either pairing similar observations together (known as agglomerative clustering) or sequentially splitting the observations (known as divisive clustering). In this chapter we will only examine agglomerative hierarchical clustering. To understand what hierarchical clustering is, let's make a small data frame based on some observations of some everyday animals:

```
lays.eggs <- c(0, 0, 1, 1, 1)
has.fur <- c(1, 1, 0, 0, 0)
breathes.air <- c(1, 1, 0, 1, 1)

animal.df <- data.frame(has.fur, lays.eggs, breathes.air)
rownames(animal.df) <- c("dog", "cat", "fish",
                         "bird", "crocodile")

animal.df

          has.fur lays.eggs breathes.air
dog             1         0            1
cat             1         0            1
fish            0         1            0
bird            0         1            1
crocodile       0         1            1
```

Notice that in this data frame we are only using 1s and 0s for our data, and we are including row names through the **rownames()** function. The row names will act as a way of labelling observations. We can use the **dist()** function to calculate a measure of pairwise similarity (or more precisely – dissimilarity) which will be displayed as a 'distance matrix' which summarises how far each observation is from another in terms of Euclidean distance (using the *method = "euclidean"* argument; see shaded box for description):

```
animal.df %>%
  dist(method = "euclidean")
```

```
                dog       cat      fish      bird
cat        0.000000
fish       1.732051 1.732051
bird       1.414214 1.414214 1.000000
crocodile  1.414214 1.414214 1.000000 0.000000
```

Notice that in our example a cat is zero distance from a dog, and a bird is zero distance from a crocodile. If we look at the `animal.df` data frame we see that this is the result of cat–dog and bird–crocodile entries being exactly the same.

What is Euclidean distance? Euclidian distance is the straight line distance between points. It is based upon Pythagoras' theorem ($h^2 = a^2 + b^2$) but extended for multiple dimensions. In practice this means adding a squared term for each new variable. We can manually calculate the Euclidian distance between a dog and crocodile from the differences between their columns for each of the three variables we recorded (`has.fur`, `lays.eggs`, and `breathes.air`):

$$dog - crocodile = \sqrt{(1-0)^2 + (0-1)^2 + (1-1)^2}$$
$$dog - crocodile = 1.414214$$

Likewise for dog and fish:

$$dog - fish = \sqrt{(1-0)^2 + (0-1)^2 + (1-0)^2}$$
$$dog - fish = 1.732051$$

We complete the analysis by using the **hclust()** function to undertake an agglomerative hierarchical clustering and then we plot it in base R. The code below results in a plot which is called a dendrogram showing the relationship between our observations in terms of (dis)similarity.

```
my.clust <- animal.df %>%
  dist(method = "euclidean") %>%
  hclust(method = "ward.D")
```

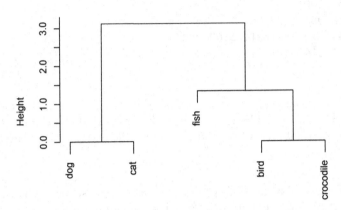

hclust (*, "ward.D")

What the hierarchical clustering process does is it finds the pair of samples that are the most similar. Once found, the most similar pairs are merged to together and then the distance matrix recalculated. This process continues in a step-wise fashion until all the samples have been merged. The y-axis (height) in the dendrogram represents the accumulative distance between the most similar pairs (calculated between mergers). However, the rules for the mergers and the calculation of their distance depends on the the the **hclust()** *method* used[2]. Regardless of method, the value of height will be stored in our cluster object as `$height`:

```
my.clust$height

[1] 0.000000 0.000000 1.333333 3.115049
```

The `my.clust$height` values can help us understand this dendrogram. In our original distance matrix as there was no difference in distances between *dog-cat* and *bird-crocodile* these were both plotted at zero on the y-axis. After the first merger fish was judged to be closest to the bird-crocodile pair at distance of 1.33. The next (and final) merger took place between fish–bird–crocodile and the dog–cat groups when the distances summed to 3.12 (which matches the height of the y-axis).

---

[2]in this example we used *method* = *"ward.D"* which is a method which tends to produce tight clusters

Therefore we read this dendrogram from the bottom upwards. The height of the dendrogram represents the order in which the clusters were joined. The ends with the labels are called 'leaves' and the lines which connect the clusters are called 'branches'. What we can see is the dog–cat and bird–crocodile clustered out first. We also see that the fish is more similar to the bird–crocodile cluster than the dog–cat cluster (a result of the later not laying eggs). Unlike other graphs the position of observation along the x-axis is not important. The cluster groupings in a dendrogram can spin on their branches and retain exactly the same meaning. In this way, even though 'cat' appears physically close to 'fish' in our dendrogram this does not mean anything.

It is important to note that the **dist()** function only works with columns that are numeric. For this reason we used used dummy variables rather than categories in `animal.df`. Many data sets we work will have variables made up of factors or characters. If we have a data set (such as `questionnaire`) made up of characters or factors we can change all of them into dummy variables using the **dummyVars()** function from the **caret** package. By using the *fullRank = TRUE* argument we ensure our dummy variables reflect the levels of our categories (in the same way as a design matrix does – see Chapter 15). Finally, to complete this process we must turn the output back into a data frame (note the use of the **predict()** function). The **dummyVars()** function will only transform columns made up of characters and factors – it will never transform numeric columns:

```
dummy <- dummyVars(" ~ .", data = questionnaire,
                   fullRank = TRUE)
dummy.ques <- data.frame(predict(dummy,
                          newdata = questionnaire))
```

Rather than using the base R plot, we can use the **fviz_dend()** function from the **factoextra** package to make easy to alter graphs (which behave like ggplots). We use the argument *show_labels = FALSE* to prevent the labels of the leaves cluttering the plot:

```
our.dendro <- dummy.ques %>%
  dist(method = "euclidean") %>%
  hclust(method = "ward.D")

fviz_dend(our.dendro, show_labels = FALSE)
```

Cluster Dendrogram

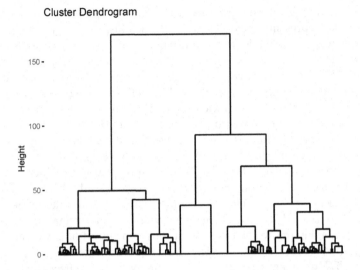

If we have a particular number of clusters in mind, we can use the **cutree()** function with the k argument to state the number of clusters. This function then assigns each observation to a cluster which we can store in a vector for later use as a variable for further modelling.

```
groups <- cutree(our.dendro, k = 2)
```

We can also easily colour the branches of a dendrogram using the **fviz_dend()** function package:

```
fviz_dend(our.dendro, k = 2, show_labels = FALSE)
```

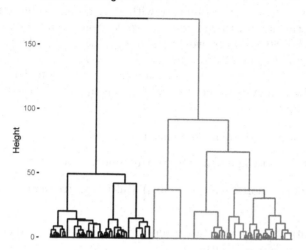

In conservation and development projects we may use the output of clustering to help guide us in how we target our interventions (e.g. the development of awareness materials). In such cases we may have a good idea of the maximum number of clusters because we have limited budget. For example, we might only have enough money to create a maximum of two different communication approaches – as a result we would know $k = 2$.

17.5.2 Dimension reduction

One of the problems with clustering is that there is no universally accepted way of determining the number of clusters needed. As a result, the final number of clusters will reflect a human decision to some extent. Additionally, clustering treats all the variables as being independent, as a result if variables are correlated then the clusters will reflect this bias. One way to remove issues involving correlation is to use dimension reduction.

Dimension reduction is the process of reducing a data set with lots of correlated (or uncorrelated) variables into uncorrelated derived variables (called dimensions) while retaining the essence of the original data. Often we will find that we can describe most of the data using just a few dimensions. In this way, dimension reduction is particularly helpful when dealing with data which has a large number of variables[3]. Questionnaires used by conservation and development projects often involve dozens of questions. Intuitively, when we start analysing questionnaire data we may like to begin searching for patterns by graphically comparing the responses to our questions against each

---

[3]usually called 'features' by computer scientists

other in pairs. The problem with this approach is the number of potential graphs is very large. A short questionnaire with 10 questions would result in 45 graphs, a questionnaire with 25 questions would result in 300 graphs, but a questionnaire with 50 questions would result in 1,225 graphs. The amount of staff hours required to build and evaluate such graphs simply isn't practical. An alternative approach is to use a use dimension reduction. The type of dimension reduction analysis we need to do depends upon the data we have. The common approaches are:

- Quantitative data – Principal Component Analysis.

- Qualitative data – (two groups only) Correspondence Analysis.

- Qualitative data – (more than two groups) Multiple Correspondence Analysis.

- Mixed quantitative and qualitative data – Multiple Factor Analysis or Factor Analysis of Mixed Data.

In the following example we will use Multiple Correspondence Analysis (MCA) to examine the data from a questionnaire designed to evaluate a project (using the `questionnaire` data set). Regardless of whether we use MCA or another approach, the steps are similar across the different methods. The number of dimensions that the analysis will return is the same as the total number of parameters. The outputs are usually called dimensions or, in the case of Principal Component Analysis, principal components. Regardless, these dimensions are always ordered by the amount of variability in the data they explain (i.e. the 1st dimension will always explain more variability than the 2nd dimension etc.). Sometimes it is possible to describe a data set using the first two dimensions. The basic steps for Multiple Correspondence Analysis are outlined below:

Step 1: run a multiple correspondence analysis with the **MCA()** function from the `FactoMineR` package:

```
mca.results <- MCA(questionnaire, graph = FALSE)
```

In the current line of code above all variables are considered in the analysis. However we have the option of using the *quanti.sup* and *quali.sup* arguments in the **MCA()** function to specify that certain quantitative (numeric) or qualitative (categorical) columns are not used for dimension reduction but are mapped on to outputs to visually assist with interpretation at a later stage. One reason to use use supplementary variables is if we are more interested in using them as a check on our outputs, rather than having them contribute to the variability in our data set.

Step 2: check the amount of variation explained by each dimension using the **get_eig()** function. The 'eig' stands for eigenvalues, in this situation these show the amount of the variation explained by each dimension:

```
get_eig(mca.results)
```

The first three lines of the output are shown below:

```
      eigenvalue variance.percent cumulative.variance.percent
Dim.1  0.3509420       22.56056                   22.56056
Dim.2  0.2324989       14.94636                   37.50691
Dim.3  0.1675925       10.77381                   48.28072
```

Often it is easier to understand the contribution of each dimension through a scree plot using the **fviz_screeplot()** function (a max of 10 dimensions are shown, but this can be changed using the *ncp* argument):

```
fviz_screeplot(mca.results, addlabels = TRUE)
```

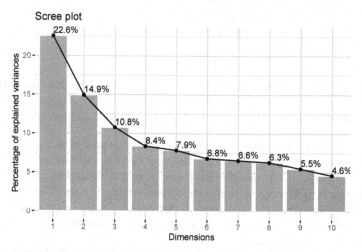

Step 3: Explore which variables and answers are the most characteristic of each dimension using the **dimdesc()** function. As the output is verbose (i.e.

long) we can we can restrict the details to just the first two dimensions by using the *axes* argument:

```
dimdesc(mca.results, axes = 1:2)
```

The first component of the output gives the r^2 associated with that particular dimension. For our first two dimensions the r^2 values were:

```
                   R2          p.value
vanilla     0.91346871 8.487900e-267
income      0.91557228 9.989817e-266
overall     0.84538873 4.569666e-199
forestry    0.27848225  3.404043e-37
trapping    0.09000229  7.389855e-12
fish.trial  0.07879618  1.658860e-10
clinic      0.01775806  2.829869e-03
watershed   0.01679601  3.696891e-03


                   R2          p.value
gender      0.73913647 2.020079e-147
clinic      0.52891120  1.964431e-83
income      0.47521765  4.411528e-69
watershed   0.14829944  4.019122e-19
overall     0.13801276  3.823694e-15
fish.trial  0.03513631  2.460465e-05
trapping    0.02004250  1.505031e-03
```

From the r^2 values of the output we see that Dim 1 is strongly correlated with the questions related to: vanilla (0.91), income (0.92), and overall (0.85), while Dim 2 is correlated with: gender (0.74), clinic (0.53), and income (0.48).

Step 4: Explore the clustering present in our observations visually through through the **fviz_mca_ind()** function. The top two dimensions will be the axes of the graphs[4]. If the points are closer together it means that these observations were more similar than other observations (i.e. in our example they would have answered the questions in our questionnaire more similarly). From the graph below we can see there looks to be two clusters:

---

[4]alternatively, we could see a similar graph by using the **MCA()** function in Step 1 with the argument *graph = TRUE*

```
fviz_mca_ind(mca.results)
```

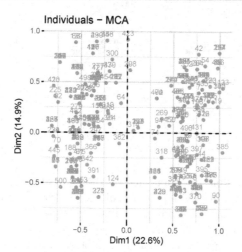

Next, we can use **fviz_mca_var()** function for the relative position of our variables. The graph is labelled with the answers to the questionnaire's questions. Answers which were often given together appear closer together.

```
fviz_mca_var(mca.results)
```

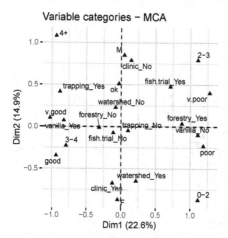

If we want both of these graphs overlaid on each other we can use **fviz_mca_biplot()** instead, which will result in:

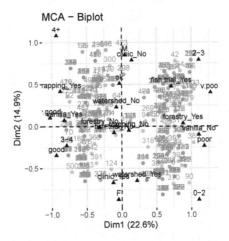

Step 5: Graph key variables (as identified by the r^2 in Step 3) to understand how they may be influencing the separation of the clusters. In order to do this we need to specify the variable from our original data set using the *habillage* argument. Additionally, this variable will need to be a factor – we can use the **as.factor()** function to accomplish this. Labels can be suppressed using the *label = "none"* argument. We can also title any of these plots using ggplot functions:

```
fviz_mca_ind(mca.results,
             habillage = as.factor(questionnaire$vanilla),
             addEllipses = TRUE, label = "none")+
  labs(title = "Clustering: vanilla")
```

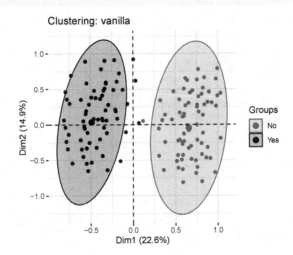

In the example above we see that `questionnaire$vanilla` category appears separate the clusters quite well.

Step 6: We can then make a dendrogram using agglomerative hierarchical clustering based on our factor analysis using the **HCPC()** function. This function will search for the optimal number of clusters based on a range of clusters we give the function using the *min* and *max* arguments[5].

```
cult.results <- HCPC(mca.results, min = 2, max = 10,
                     graph = FALSE)
```

The function will recommend the number of clusters to split the data into through a line on a dendrogram. We can also look at the dendrogram through:

```
fviz_dend(cult.results, show_labels = FALSE)
```

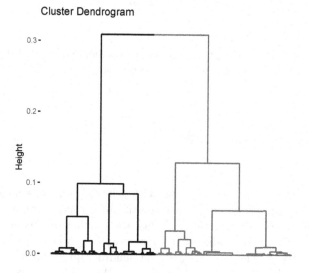

Cluster Dendrogram

Alternatively, we can look at how clusters have been assigned on a dimension plot:

---

[5]the function bases this split on a statistical property called inertia

```
fviz_cluster(cult.results)
```

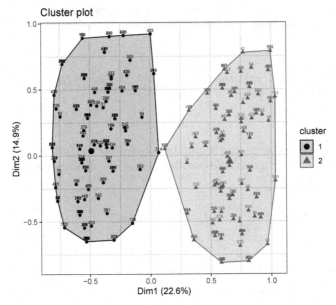

If we are not prepared to accept the clusters recommended by the **HCPC()** function we can run the function with the argument *graph = TRUE* (graph not shown):

```
HCPC(mca.results,  graph = TRUE)
```

This will produce an interactive graph with a line indicating a suggested split on the dendrogram. We can either click on the line to accept the suggested split or click upwards or downwards to change the number of clusters. After doing this the function will produce two more graphs: a 3-D graph showing the relationship between the clusters and the dendrogram and a factor map showing the clusters in relation to our initial plot. We can also make the 3-D plot directly by using the *argument choice = "3D.map"*:

```
plot(cult.results, choice = "3D.map")
```

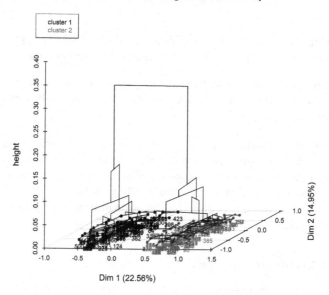

Step 7: We can then examine the cluster assigned to each observation stored within the `data.clust` column in the object made by **HCPC()** :

```
cult.results$data.clust
```

We can then bind this output with our original data so that we can undertake further exploration:

```
data.clusters <- cbind(questionnaire, cult.results$data.clust)
```

17.6 Classification

Classification is a name for statistical techniques which can predict (classify) the category of a response variable based on a set of observations. We could use these methods in conservation and development for:

- Predicting whether a household would want to participate in a health, agriculture, or conservation program based on data collected on aspects of their everyday lives.

- Predicting the vulnerability of wildlife populations and protected areas to illegal activities based on geographic and socio-economic data.

- Predicting participant satisfaction with a conservation or development project (and understanding how those perceptions might be formed) using questionnaire data.

There are a large and growing number of classification techniques. Some of the simpler techniques include recursive partitioning (also known as decision trees), random forests, and naive Bayes. Indeed, logistic regression and cummulative link models (which we covered in Chapter 16) can also be regarded as a type of classification[6]. In this chapter we will introduce just one type of classification technique: classification trees.

17.6.1   Classification trees

Many classification techniques are based on what are known as tree-based methods. These use techniques use splitting rules to split the data into categories. These splits can be thought of as decisions. As a result, these are often known as 'decision trees'. Decision trees can be used for both regression and classification (sometimes jointly referred to as Classification And Regression Trees, or CART, for short). For purposes of this chapter we are only looking at classification trees.

Many modern classification techniques are amazing at predicting categorical response variables accurately, however, they can be very hard to interpret. However, in conservation and development we usually want to understand and replicate the conditions which lead to success. Consequently, we are usually interested in how explanatory variables contribute to an outcome (not just an accurate prediction). In this way classification trees have a big advantage – usually they are very easy to explain, can be understood by everyday people, and often appear to mirror human decision making. Additionally classification trees are a non-parametric approach and consequently they do not have to conform to a particular statistical distribution – and therefore carry fewer assumptions than parametric approaches. The downside is that classification trees usually have less predictive power than other classification methods. However, as classification trees are often the stepping stone to learning other tree-based methods they are usually a good starting point for learning about classification.

---

[6]as logistic regression predicts a two-class binary response (yes or no, true or false, alive or dead), and cummulative link models predict ordinal categories

17.6.2 How classification trees work

Classification trees result in a graphical output showing how the classification was made. Let's imagine doing an analysis to predict when people take umbrellas, wear raincoats, or normal clothes. This might result in a classification tree, like the one in Figure 17.1. Classification trees have similarities to flow-charts. We begin at the top of the figure at a test for the first split (called a root node). Depending on the 'yes' or 'no' outcome of the test we flow downwards along branches (the outcome of the test) into either more nodes (which have a logical test) or to a leaf (sometimes called a terminal node) which predicts the final category. Using such a tree we are able to assign any row of observations from our data into a category.

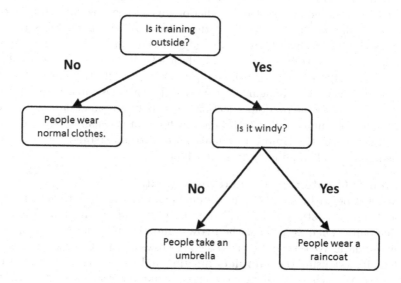

FIGURE 17.1

A decision tree predicting when people will take normal clothes, a raincoat, or carry an umbrella

Classification trees work by predicting the category of a response variable by finding a rule to split the data at each node which results in the split having the highest purity (i.e. the lowest possible rate of misclassification)[7]. The splitting process then continues. One of the problems with this step-wise method is it cannot assess, in advance, the series of splits that would

---

[7]in classification trees this is normally measured by the Gini index

be required for the leaves to have the highest possible purity. As a result, trees frequently becomes deeper (having more splits) than is optimal. Trees which are very deep typically have a problem with over-fitting (i.e. being overly specific to our data – see Chapter 14). One way around this is to use a complexity parameter[8] which penalises the model according to the number of nodes. The classification tree can then be pruned (shortened) on the basis of the complexity parameter to a simpler model that is less likely to be over-fitted.

17.6.3   Model evaluation

The standard model evaluation procedure in machine learning is to randomly split data sets into training and testing data sets. This is known as cross-validation. The idea is to first build a model using the training data set and then test the accuracy of its predictions using the unseen testing data. The reason why we do not use all of the data to build our model is that it will normally lead to over-fitting. In large data sets you may be able to split your data 50:50 into training and testing data sets. If your data set is smaller then you may choose to have a larger proportion in the training data set (e.g. 70:30). It is easy to imagine that quite different trees could be produced with a slightly different random selection (especially for small data sets). For this reason a technique called k-fold cross-validation is usually preferred which we will cover shortly. However, it makes sense to understand the basic method using an equal training–testing split first:

Step 1: Create training and testing data sets
We will begin by simply splitting our **questionnaire** data set into training and testing data sets. We will start by using **set.seed()** function for reproducibility. Then we will use the **nrow()** function to count the number of rows in our data. We then randomly assign the rows of **questionnaire** to be part of training set using the **sample()** function and make this into an object called index. We then subset **questionnaire**: making rows identified by index part of **training** data set, and the rest part of the **testing** data set:

```
set.seed(14)
no.train <- 0.5 * nrow(questionnaire)
index <- sample(nrow(questionnaire), no.train, replace = FALSE)
training <- questionnaire[index, ]
testing <- questionnaire[-index, ]
```

Step 2: Build the model

---

[8]also known as a cost-complexity factor

We then make a model using an additive model formula including all the variables of interest using the **rpart()** function (short for recursive partitioning) and name the training data set. Because **rpart()** can also be used for a regression tree we need to use the argument *method = "class"*. You should double check that `rpart` and `rpart.plot` packages are loaded, otherwise the function will not work.

```
our.tree <- rpart(overall ~ vanilla + gender + clinic + income,
                  method = "class", data = training)
```

Step 3: Examine the model and its predictions
We can use the **prp()** function to see the tree created by the model. By using the *type = 5* argument we get a plot with intuitive labels:

```
prp(our.tree, type = 5)
```

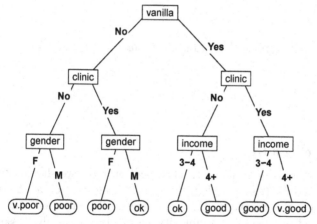

Next we can make predictions based on our model:

```
our.pred <- predict(our.tree, testing, type = "class")
```

We can then use this the model output to check whether or not the model correctly classified the response variable. We do this by using the **table()** function (notice how we can label the output). The table that is produced is called a confusion matrix. A confusion matrix shows the frequency with which

categories were correctly classified or misclassified by the model. Categories
which were correctly classified appear on the diagonal.

```
conf.matrix1 <- table(prediction = our.pred,
                      actual = testing$overall)
conf.matrix1

          actual
prediction v.poor poor ok good v.good
   v.poor    11    4   0    0     0
   poor       9   54  12    0     0
   ok         0    3  50   14     0
   good       0    0   7   72     4
   v.good     0    0   0    3     7
```

We can use the proportion of categories correctly classified on the diagonal
as a measure of our model's accuracy using the **diag()** function in combination
with the **sum()** function. This shows this model has an accuracy rate of ~78%:

```
accuracy<- sum(diag(conf.matrix1)) / sum(conf.matrix1)
accuracy

[1] 0.776
```

17.6.3.1   Pruning a tree

As there is a risk of over-fitting with classification trees we should trial the
effectiveness of shorter (less complex) trees. To do this we look at the output
of the model using the **printcp()** function, identify the value with the lowest
xerror value and identify the complexity parameter (cp) value associated
with it. Next, we shorten (prune) the the tree using the complexity parameter
we have identified. In this process we are trying to see if a shorter tree can
give us comparable results:

```
printcp(our.tree)
```

```
Classification tree:
rpart(formula = overall ~ vanilla + gender + clinic + income,
    data = training, method = "class")

Variables actually used in tree construction:
[1] clinic  gender  income  vanilla

Root node error: 167/250 = 0.668

n= 250

        CP nsplit rel error   xerror       xstd
1 0.371257      0   1.00000  1.00000  0.044587
2 0.107784      1   0.62874  0.62874  0.046730
3 0.041916      2   0.52096  0.52096  0.045099
4 0.035928      3   0.47904  0.51497  0.044976
5 0.023952      4   0.44311  0.49102  0.044450
6 0.011976      6   0.39521  0.43114  0.042874
7 0.010000      7   0.38323  0.43713  0.043049
```

Based on the output we see that lowest **xerror** term occurs when CP is 0.011976 and 6 splits have been made. We can then shorten the tree using the **prune()** function with this **cp** value as an argument.

```
pruned.model <- prune(our.tree, our.tree$cp[6])
```

We then examine the simplified tree produced:

```
prp(pruned.model, type = 5)
```

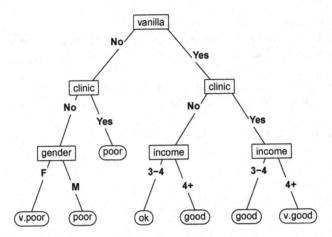

We then make a new set of predictions:

```
pruned.pred <- predict(pruned.model,testing, type = "class")
```

And repeat the evaluation process by examining the confusion matrix and model accuracy:

```
conf.matrix2 <- table(pred = pruned.pred, true = testing$overall)
conf.matrix2

          true
pred      v.poor poor ok good v.good
  v.poor      11    4  0    0      0
  poor         9   57 24    0      0
  ok           0    0 38   14      0
  good         0    0  7   72      4
  v.good       0    0  0    3      7
```

```
accuracy.short <- sum(diag(conf.matrix2)) / sum(conf.matrix2)
accuracy.short

[1] 0.74
```

What we see is that the accuracy has dropped to 0.74, only slightly lower than the ~0.78 we previously recorded in our initial model. On this basis we would decide that our new model might be the better of the two because although it is a little bit less accurate it is probably less prone to over-fitting.

17.6.4 *k*-fold cross-validation

An alternative to an equal training and testing data split is to use what is known as *k*-fold cross-validation. In *k*-fold cross-validation your data set is randomly divided into a number (*k*) of samples ('folds'). One of these folds is held back as a test data set while the rest are used for training. The process is then repeated *k* times (the iterations) with a different fold being used as the testing data each time (Figure 17.2). As there are now multiple results these will be used to optimise the depth of the tree to prevent over-fitting.

	Fold 1	Fold 2	Fold 3	Fold 4	Fold 5
Iteration 1	Train	Test	Test	Test	Test
Iteration 2	Test	Train	Test	Test	Test
Iteration 3	Test	Test	Train	Test	Test
Iteration 4	Test	Test	Test	Train	Test
Iteration 5	Test	Test	Test	Test	Train

FIGURE 17.2
Visualising 5-fold cross-validation

The **caret** package (short for classification and regression training) makes *k*-fold cross-validation easy. First, we use the **traincontrol()** function to make an object which contains the number of data splits (*number = 5*) for cross-validation (*method = "cv"*). Then this object is then passed to the **train()** function using the *trControl* argument along with the argument for the technique we are using (*method = "rpart"*):

```
set.seed(27)
cross.val <- trainControl(method = "cv", number = 5)
train.rpart <- train(overall ~ vanilla + gender + clinic +income,
                     data = questionnaire,
                     method = "rpart",
                     trControl = cross.val)
```

When we run the model object (`train.rpart`) a summary of the average results from the data splits is produced, including the accuracy for different complexity parameters:

```
train.rpart
```

```
CART

500 samples
  4 predictor
  5 classes: 'v.poor', 'poor', 'ok', 'good', 'v.good'

No pre-processing
Resampling: Cross-Validated (5 fold)
Summary of sample sizes: 401, 399, 402, 399, 399
Resampling results across tuning parameters:

  cp          Accuracy   Kappa
  0.03963415  0.6600264  0.5250652
  0.10670732  0.6117767  0.4505394
  0.37500000  0.4420050  0.1631796

Accuracy was used to select the optimal model using the
 largest value.
The final value used for the model was cp = 0.03963415.
```

We can extract the information for graphing the from the model object using `$finalModel`:

```
prp(train.rpart$finalModel, type = 5)
```

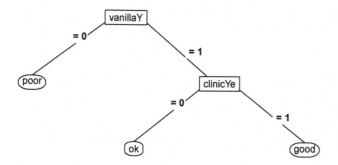

What we see is that rather than showing 'no' and 'yes' in our splits, **caret** has handled these as dummy variables (changing them to '0' or '1'). This makes interpreting these trees a little more complex. Care should be taken to carefully read the nodes as the branches reflect a negative ('0') or positive response ('1') to the question attached to the variable in the node (e.g. clinicYe is short for Clinic = 'Yes'). Our outcome is now only supporting a relatively short tree which has splits based on involvement with the vanilla program and clinic attendance. The accuracy we now have (0.66) is lower than what we obtained initially from our trial with **rpart()**. Despite this we can be more confident in the reliability of this shorter tree because it was the outcome of multiple trials.

Aside from the ability to do k-fold cross-validation, one of the great advantages of using the **caret** package is the ability to evaluate different machine learning modelling approaches against each other using cross-validation (**caret** current supports over 200 different approaches). For example, we could test a random forest model (a different type of tree-based machine learning method) by simply changing the argument to *method = "rf"*. While we don't have to load the **randomForest** package directly in the code (because **caret** will automatically find it) it does need to be previously installed. Random forests makes predictions based a large number of short trees built from randomly selected variables where 'mtry' represents the number of variables randomly sampled at each split. A comparison by accuracy shows that the random forest model has slightly better accuracy than our original classification tree (the trade off with this improved accuracy is that the random forest model cannot be graphed as a tree and so is less interpretable):

```
set.seed(51)
train.rf <- train(overall ~ vanilla + gender + clinic + income,
                  data = questionnaire,
                  method = "rf",
                  trControl = cross.val)
train.rf
```

Random Forest

500 samples
  4 predictor
  5 classes: 'v.poor', 'poor', 'ok', 'good', 'v.good'

No pre-processing
Resampling: Cross-Validated (5 fold)
Summary of sample sizes: 400, 400, 400, 399, 401
Resampling results across tuning parameters:

mtry	Accuracy	Kappa
2	0.7359904	0.6387496
4	0.7679310	0.6822279
6	0.7639310	0.6770086

Accuracy was used to select the optimal model using the
 largest value.
The final value used for the model was mtry = 4.

17.6.5 Accuracy versus interpretability

Before we choose a classification method (or a set of methods) we need to think about whether we are more interested in interpretability or accuracy. If we want to understand why an intervention worked and want to replicate the outcome in the future we are probably more interested in interpretability and may choose a technique like classification trees. Alternatively, if it is critical that predictions be accurate then classification trees are unlikely to be the best method, rather we would need to explore other more advanced classification methods (such as those described in Applied Predictive Modeling (2013) by Max Kuhn and Kjell Johnson).

17.7 Recommended resources

- Gareth James, Daniela Witten, Trevor Hatsie, and Robert Tibshirani (2013). An Introduction to Statistical Learning: with Applications in R. Springer, London.

- Max Kuhn and Kjell Johnson (2013). Applied Predictive Modeling. Springer, London.

- `https://topepo.github.io/caret`. The online guide to the `caret` package by Max Kuhn.

17.8 Summary

- Clustering is a form of unsupervised learning used to understand hidden patterns or relationships within data.

- Dimension reduction is a useful approach for simplifying a large number of variables into a smaller set of derived variables.

- Classification is a name for the statistical techniques that predict the category of a response variable.

- Classification trees are the basis for many classification techniques, they are also very easy to interpret and therefore are an entry point for understanding more complex classification techniques.

- The standard approach for evaluating classification models is through cross-validation.

18

Reporting and worked examples

In this chapter we demonstrate:

- How to incorporate the results of a data analysis into a project report.

- How to discuss possible follow-up actions.

18.1 Writing the project report

Different organisations and donors have different reporting formats. Generally, most reports consist of three major sections: (1) an introduction, (2) a section summarising the activities, and (3) a discussion on the effectiveness of the project as a whole. It is in the second section that we incorporate the results of our data analysis. Usually, we are asked to write a single paragraph summarising each project activity. Even though our analysis of the data may have been complicated we need to communicate simply. Below are some useful tips for writing an activity summary:

- Use plain language. Readers of the report don't want to read statistical jargon.

- When reporting on an activity write it like a short story with a:

  - Beginning: outline what happened in the activity.
  - Middle: describe the results of the activity in terms of the project indicators you were monitoring.
  - End: briefly evaluate the results in terms of the project's expectations. If the activities' underlying assumptions were wrong – describe why and what it meant for this project, as well as what it may mean for the design of future projects.

As you may have noticed, the summary will refer back to information outlined in the project's logframe (i.e. the activity, its indicators, and its assumptions). For this reason, throughout the project you should be checking that the project aligns with the logframe. If it becomes clear that activities cannot be undertaken, or monitored in the way planned (e.g. certain indicators may turn out to be impractical) it is important to get in touch with the donor so that changes can be agreed upon.

Some reporting formats allow for more detailed information to be included at the end of the reports (usually called annexes, appendices, or supplementary material). This is where more detailed documentation of statistical methods, results (e.g. graphs, test statistics, and model selection tables), and conclusions can be put.

Reports may or may not include a section on recommendations or follow-up actions. Regardless, of whether or not these are asked for in the project report — it is in the organisation's best interests to document and discuss these so it can adapt and refine its future projects. This last step represents the beginning of the Act phase of the Deming cycle (see Chapter 3).

In the boxed sections which follow I demonstrate three activity summaries and align each of these with their respective logframe. The first example emphasises data wrangling, and graphing, the second deals with count data and a simple, before-after study design, while the third deals with a more complex, before-after control-impact, study design. I also include a section called 'Follow-up action' — this represents information for a section in the project report or internal discussion within the organisation about future approaches. Each section finishes with the R code (and outputs) that formed the basis for the activity summary.

18.2   The packages

A number of packages are required for each of the demonstration analyses. These can be found at the start of the code that follows each section.

18.3   The data

For this chapter we will use a number of example data sets from the condev package.

```
library("condev")
data(attendees)
data(prosecutions)
data(fishponds)
```

18.4   Reporting training data

When donors request performance information on activities which relate to training they normally require information on dates and a break down of the gender of the people attending. This process is known as disaggregating data. Most donors will ask for a summary of attendance data disaggregrated by gender. Often we will attempt to achieve a specific gender ratio in training. Consequently, we can use chi-squared tests to see if we achieved our target. While some donor staff will be familiar with chi-squared testing many will not, consequently, graphs will be better for interpretation.

Logframe:

- **Activity:** village livelihood training courses on vanilla farming, sustainable forestry, and watershed management were undertaken.

- **Assumption:** a female participation rate of 40% will be achievable.

- **Indicator**: number and gender of people attending training sessions.

- **Data source**: attendance sheets: `data(attendees)`.

Activity summary:

Three training courses on vanilla farming, sustainable forestry, and watershed protection were run across 5 villages (Anota, Lamaris, Maniwavie, Nugi, and Takendu) in October 2018. The total numbers of attendees for each course is given below:

```
  training.dates forestry vanilla watersheds
1     2018-10-05       92       -          -
2     2018-10-08        -      95          -
3     2018-10-12        -      47          -
4     2018-10-06        -       -        100
5         Total       92     142        100
```

On average, each participant attended two training sessions (regardless of gender). A total of 156 individuals (59 females and 97 males) actually attended the training courses. The observed female participation rate (approximately 38%) was broadly consistent with our 40% target. The gender ratio was consistent across all three training courses (Figure A).

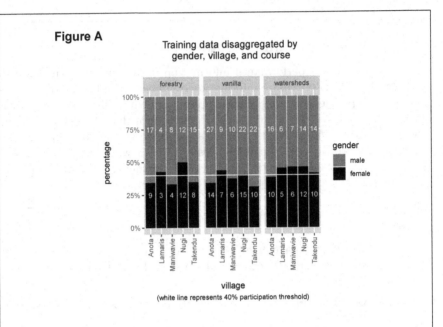

Figure A

Training data disaggregated by
gender, village, and course

Follow-up action:
From our analysis of the attendance data it is clear that a female participation rate of 40% is achievable. However, achieving gender parity is unlikely in the near term without a program in place to enable greater attendance by women and girls (e.g. ensuring training is timed to allow female attendance and child minding is provided for attendees). It was clear from the popularity of the vanilla farming course that there is a great deal of interest in developing cash crops on the island. Consequently, the potential of other conservation compatible cash crops for the island should be explored.

Supporting R code and key outputs:

```
library("janitor")      # for adding row and column totals
library("lubridate")    # for simplifying dates
library("scales")       # for percentages on ggplot axis
library("tidyverse")    # for graphing and data wrangling

# data
library("condev")
data(attendees)
```

```
# number of sessions attended by each individual
info <- attendees %>% group_by(gender, name)%>%
  summarise(sessions = n())

# number of individuals
gender.ratio <- table(info$gender)
gender.ratio

female   male
   59     97

# consistent with 40% female participation rate?
# null hypothesis = 40:60 gender ratio
# alternative hypothesis = different gender ratio
chisq.test(gender.ratio, p = c(0.4, 0.6))

      Chi-squared test for given probabilities

data:  gender.ratio
X-squared = 0.30876, df = 1, p-value = 0.5784

# test for stable gender ratio (i.e. test of independence)
chisq.test(attendees$gender, attendees$course)

      Pearson's Chi-squared test

data:  attendees$gender and attendees$course
X-squared = 1.0025, df = 2, p-value = 0.6058

# average number of sessions attended
# because NAs present argument na.rm must be used
info %>%
  pivot_wider(names_from = gender, values_from = sessions) %>%
  summarise(F.avg = mean(female, na.rm = TRUE),
            M.avg = mean(male, na.rm = TRUE))

# A tibble: 1 x 2
  F.avg M.avg
  <dbl> <dbl>
1  2.22  2.09

# use the lubridate function to read dates
attendees$training.dates <- dmy(attendees$training.dates)

# we want "female" on the base of a stacked graph so
# we need to relevel the gender variable
attendees <- attendees %>%
  mutate(gender = fct_relevel(gender, "male", "female"))
```

```
# make a data set containing the count information
# using count=n()
trainees <- attendees %>%
  group_by(course, gender ,village) %>%
  summarise(count = n())

# produce a table of training dates
training.days <- attendees %>%
  group_by(course, training.dates) %>%
  summarise(count = n_distinct(name)) %>%
  pivot_wider(names_from = course , values_from = count) %>%
  adorn_totals("row")

# replace NAs with '-'
# need to change to data frame to work
training.days %>%
  as.data.frame() %>%
  replace(is.na(.), "-")
```

```
  training.dates forestry vanilla watersheds
1     2018-10-05       92       -          -
2     2018-10-08        -      95          -
3     2018-10-12        -      47          -
4     2018-10-06        -       -        100
5          Total       92     142        100
```

```
# critical graph
# Note: all labels (e.g title) should appear on a single line -
# their split appearance is due to the book's margins.
# In geom_text() we use position = position_stack() to make
# the text follow the divisions within the stacked bar graph

ggplot()+
  geom_bar(data = attendees, aes(x = village, fill = gender),
           position = "fill")+
  geom_hline(yintercept = 0.4, colour = "white")+
  geom_text(data=trainees, aes(x = village, y = 0.5,
                               label= count,
                               group = gender),
            size = 3, position = position_stack(vjust = 0.5),
            colour = "white")+
  facet_grid(.~course)+
  scale_fill_manual(values = c("grey50", "black"))+
  scale_y_continuous(labels = percent)+
  labs(y = "percentage\n",
```

```
      x = "\nvillage",
tag = "Figure A",
title = "Training data disaggregated by \ngender, village,
and course\n",
caption = "(white line represents 40% participation
threshold)")+
theme(axis.text.x = element_text(angle = 90, vjust = 0.5,
                                 hjust = 1),
        plot.title = element_text(hjust = 0.5),
        plot.tag = element_text(face = 2, size = 16),
        plot.caption = element_text(hjust = 0.5))
```

18.5   Prosecution results

In the example below we use a model selection approach with Poisson regression to analyse a simple before-after study design.

Logframe:

- **Activity:** village magistrates from 5 villages were given environmental awareness training.

- **Assumption:** training of village magistrates will result in increased number of prosecutions regarding infringement of village environmental laws.

- **Indicator:** the change in monthly prosecutions (comparing 12 months prior to training to the 12 months after training).

- **Data source:** village court records: `data(prosecutions)`.

Activity summary:

Village magistrates in five project villages were given awareness training on the importance of enforcing environmental laws. Village court records for the subsequent 12 months were compared against the previous 12 months for evidence of any changes in prosecutions (regarding environmental laws). Observed prosecution rates doubled after magistrates were given training (Figure B) from an average of 0.95 prosecutions per village per month to 2.0 prosecutions (total annual prosecutions: before training = 57, after training = 121). Our modelling suggests that the observed increase in the number of prosecutions was a result of the training and there were no differences between villages.

Figure B

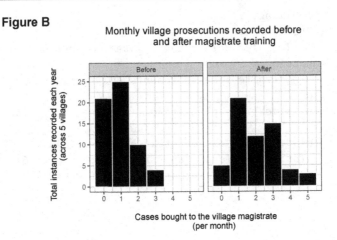

Monthly village prosecutions recorded before
and after magistrate training

Cases bought to the village magistrate
(per month)

Follow-up action:
Given the doubling of the prosecution rate, magistrate training appears
to be highly effective and should be incorporated into other projects.
Given that we do not know how long this effect will persist, follow up
monitoring should be conducted. Such monitoring would enable us to
assess how often refresher training should be given. In addition, should
a funding opportunity arise, we should undertake a before-after control-
impact study with new communities to remove the possibility of the effect
being a consequence of the year of monitoring.

Supporting R code and key outputs:

```
library("AICcmodavg")    # for model selection
library("DHARMa")        # for simulateResiduals()
library("tidyverse")     # for graphing and data wrangling

# data
library("condev")
data(prosecutions)

# candidate model list
pros.mod <- list()
pros.mod[[1]] <-  glm(cases ~ 1, family = "poisson",
                      data = prosecutions)
pros.mod[[2]] <-  glm(cases ~ village, family = "poisson",
                      data = prosecutions)
pros.mod[[3]] <-  glm(cases ~ intervention, family = "poisson",
                      data = prosecutions)
```

```
pros.mod[[4]] <- glm(cases ~ intervention + village,
                     family = "poisson",
                     data = prosecutions)

# Generate AICc table
Modnames <- paste("mod", 1:length(pros.mod), sep = " ")
aictab(cand.set = pros.mod, modnames = Modnames, sort = TRUE)
```

Model selection based on AICc:

	K	AICc	Delta_AICc	AICcWt	Cum.Wt	LL
mod 3	2	348.93	0.00	0.97	0.97	-172.42
mod 4	6	356.02	7.08	0.03	1.00	-171.64
mod 1	1	370.40	21.47	0.00	1.00	-184.18
mod 2	5	377.34	28.40	0.00	1.00	-183.40

```
# data summary
prosecutions %>%
  group_by(intervention)%>%
  summarise(average = mean(cases), total = sum(cases))
```

```
# A tibble: 2 x 3
  intervention average total
  <fct>          <dbl> <int>
1 Before          0.95    57
2 After           2.02   121
```

```
# check residuals - assumptions appear okay
# use seed to replicate results
set.seed(29)
mod.res <- simulateResiduals(pros.mod[[3]])
plot(mod.res)

# check overdispersion: okay
testDispersion(mod.res)

# check zero inflation: okay
testZeroInflation(mod.res)

# critical graph
ggplot()+
  geom_histogram(data = prosecutions, aes(x = cases),
                 binwidth = 1, colour = "white",
                 fill = "black")+
  facet_grid(.~intervention)+
```

```
scale_x_continuous(breaks = seq(from = 0, to = 10, by =1 ))+
labs(title = "Monthly village prosecutions recorded before
      and after magistrate training\n",
      tag = "Figure B",
      x = "\nCases bought to the village magistrate
      (per month)",
      y = "Total instances recorded each year
      (across 5 villages)\n")+
theme_bw()+
theme(plot.title = element_text(hjust = 0.5),
      plot.tag = element_text(face = 2, size = 16))
```

18.6   Fish pond study

In the following example, we use a model selection approach with linear mixed effect models to analyse a more complex before-after control-impact design.

Logframe:

- **Activity:** trial of high vs low rice bran feed in fish ponds.
- **Assumption:** high rice bran fish feed will increase the weight of fish harvested.
- **Indicator**: total weight of fish harvested annually.
- **Data source**: household diaries: data(fishponds).

Activity summary:
A randomised before-after control-impact trial was conducted with 12 prospective fish farmers to assess the impact of two different local fish food formulations. Each farmer was helped to construct a 100 m^2 pond and given a kilogram of starter stock. The study was undertaken for two years. Farmers recorded their harvest in diaries provided by the project. In the first year all farmers fed their fish a low bran fish food. In the second year a randomly drawn subset of 6 farmers fed their fish a high rice bran fish food while the others applied the original low rice bran fish feed.

Our statistical modelling showed that the type of fish feed did not affect the harvest. On average 27.4 kg of fish was harvested in the first year while there was an additional 6.8 kg of fish produced in the second year

(Figure C). At this stage it is not clear why the production in the second year was higher. We speculate it may be due to the fish ponds maturing over time with more aquatic larvae being present in the second year. The presence of a greater number of prey items in the second year might be responsible for the increase in harvest. Interestingly, our modelling also suggested that if there was a difference between feed types – the low rice bran feed might be marginally better (potentially producing an extra 2.2 kg of fish per year more). So we conclude, contrary to our expectation, that the high rice bran feed is not superior to the low rice bran feed.

Figure C

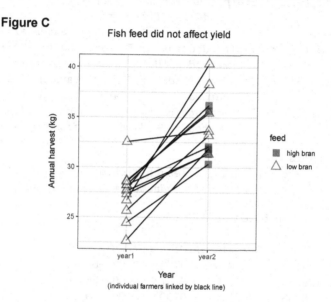

Follow-up action:
Based on the result of this trial and the large additional effort required to produce a high rice bran fish food we recommend village fish farmers should continue with their original low rice bran feed. Given the low cost of monitoring (using household diaries) ongoing community monitoring of the fishponds should be encouraged. Over time, this will allow a better understanding of fishpond production in relation to pond age.

Supporting R code and key outputs:

```
library("AICcmodavg")    # for model selection
library("car")           # for multicolinearity
library("DHARMa")        # for simulateResiduals()
```

```
library("nlme")         # for repeated measures
library("tidyverse")    # for graphing and data wrangling

# data
library("condev")
data(fishponds)

# candidate models
fish.mod <- list()
fish.mod[[1]] <- lme(production.kg ~ year + feed,
                  random = ~1|farmer,
                  method = "ML", data = fishponds)
fish.mod[[2]] <- lme(production.kg ~ year, random = ~1|farmer,
                  method = "ML", data = fishponds)
fish.mod[[3]] <- lme(production.kg ~ feed, random = ~1|farmer,
                  method = "ML", data = fishponds)
fish.mod[[4]] <- lme(production.kg ~ 1, random = ~1|farmer,
                  method = "ML", data = fishponds)

# AICc table
Modnames <- paste("mod", 1:length(fish.mod), sep = " ")
aictab(cand.set = fish.mod, modnames = Modnames, sort = TRUE)
```

Model selection based on AICc:

```
      K    AICc  Delta_AICc  AICcWt  Cum.Wt      LL
mod 2 4  124.46        0.00    0.63    0.63  -57.18
mod 1 5  125.53        1.07    0.37    1.00  -56.10
mod 4 3  145.53       21.07    0.00    1.00  -69.16
mod 3 4  146.32       21.86    0.00    1.00  -68.11
```

```
# check model outputs of top model
fish.mod[[2]]
```

Linear mixed-effects model fit by maximum likelihood
  Data: fishponds
  Log-likelihood: -57.17682
  Fixed: production.kg ~ year
(Intercept)   yearyear2
  27.370287   6.799765

Random effects:
 Formula: ~1 | farmer
        (Intercept) Residual
StdDev: 1.323484 2.310008

```
Number of Observations: 24
Number of Groups: 12

# check model outputs of competing model
fish.mod[[1]]

Linear mixed-effects model fit by maximum likelihood
  Data: fishponds
  Log-likelihood: -56.09586
  Fixed: production.kg ~ year + feed
 (Intercept)    yearyear2 feedlow bran
   25.123486     7.923165     2.246800

Random effects:
 Formula: ~1 | farmer
        (Intercept) Residual
StdDev:   0.9303434 2.338834

Number of Observations: 24
Number of Groups: 12

# check for multicolinearity issues
# using model with most parameters - vif < 5 therefore ok
vif(fish.mod[[1]])

     year      feed
 1.568307  1.568307

# check residuals of top model - assumptions appear ok
plot(fish.mod[[2]])

# alternatively we could use the DHARMa package but as it
# doesn't support nlme models we have rewrite the top model
# formula for the use by the lme4 package using lmer()
# in lmer() random effects are bracketed:

library("lme4")

# use seed to replicate results
set.seed(29)
top.mod <- lmer(production.kg ~ year + (1|farmer),
                REML = FALSE, data = fishponds)
fish.residuals <- simulateResiduals(top.mod)
plot(fish.residuals)
```

```
# critical graph
ggplot()+
  geom_path(data=fishponds, aes(x = year,
                                y = production.kg,
                                group = farmer))+
  geom_point(data=fishponds, aes(x = year,
                                 y = production.kg,
                                 fill = feed, shape = feed),
             size = 4, colour = "black", alpha = 0.5)+
labs(title = "Fish feed did not affect yield\n",
     caption = "(individual farmers linked by black line)",
     x = "\nYear",
     y = "Annual harvest (kg)\n",
     tag = "Figure C")+
scale_fill_manual(values = c("black", "white"))+
scale_shape_manual(values = c(22, 24))+
theme_bw()+
theme(plot.title = element_text(hjust = 0.5),
      plot.caption = element_text(hjust = 0.5),
      plot.tag = element_text(face = 2, size = 16))
```

19

Epilogue

Over the course of the book, you have hopefully gained some knowledge of data science tools, and a familiarity with R. You now should be in a position to apply these tools to your conservation and development projects. However, you need to be careful with your application of these tools. There is a common saying:

'If your only tool is a hammer – every problem looks like a nail'

Just because we have a tool, like R, it doesn't mean it is appropriate to every problem faced by us. It is important to understand that there is no piece of software that works for every situation. Rather, the tests and tools you have been exposed to in this book have been designed to give you confidence to begin exploring the world of data science and statistics.

New tools and packages are constantly being developed for R. For the six years I worked in Papua New Guinea, I rarely ran to another person involved in data science, and internet was patchy at best. Yet I managed to feel like I was keeping up with the R tools thanks to Twitter[1]. By following just a few twitter accounts related to my interests in R, I got small manageable snippets of information which exposed me to new R tools and helpful tips. If I couldn't figure out how to do something on my own, I would go to Stackoverflow[2] for help. Usually, I would find someone else had already asked a similar question, and a solution close to what I needed would be already waiting for me. Nowadays, face-to-face help is closer than ever with a growing network of R user groups in many towns and cities around world. Among these is a user group, called R-Ladies, which exists to promote gender diversity in the R community and is now present in over 50 countries[3].

I will end with a call to action. The life cycle of data science cannot end with the findings of the project being written into a report that is promptly forgotten about. When we are planning a project we need to think about who the key decision makers are, what their decision making process is, and develop a strategy to ensure our key findings help inform their decision making. Evidence-based decision making depends upon decision makers having access to, and understanding evidence — and it is our job to find a way to make that happen.

---

[1] https://twitter.com
[2] https://stackoverflow.com
[3] https://rladies.org

A

Appendix: step-wise statistical calculations

This appendix demonstrates how you can build your own code to calculate some of the most common statistics associated with linear modelling. For those of you who wish to have a better understanding of statistics, I recommend you practice translating standard statistical equations into R. Not only will this help you understand what these functions are doing, but it will demystify mathematical equations. While most people put their trust in R's functions the text just below the version information in the console says: 'R is a free software and comes with ABSOLUTELY NO WARRANTY'. By using R we take on board the risk that there could be a bug which might result in an incorrect result. By being able to turn mathematical equations into code we can independently check our calculations against the results R gives us.

A.1 How to approach an equation

Mathematical notation can seem daunting. To translate an equation into R we need to:

- **Step 1:** Identify the variables/objects required by the equation.

- **Step 2:** Create any new variables/objects required.

- **Step 3:** Code the equation into manageable steps.

A.2 Data

Let's construct a small data frame:

```
y <- c(3, 4, 7, 5, 8, 8, 12, 13)
x <- c(2, 4, 6, 7, 9, 10, 11, 12)
df <- data.frame(x, y)
```

A.3   The standard deviation

The equation for the standard deviation is:

$$sd = \sqrt{\frac{\sum (y - \bar{y})^2}{n - 1}}$$

If you are not familiar mathematical symbols you need to know is that the symbol $\sum$ means 'sum' and the $\sqrt{}$ symbol means 'square root'. The bar symbol on top of a variable is a symbol for 'mean', so $\bar{y}$ would be the mean of y.

Step 1: Identify the variables/objects: y, $\bar{y}$, and n.

Step 2: Create any new variables/objects: none required.

Step 3: Code the instructions into manageable steps: as $\bar{y} = $ mean(y) and n = length(y) we can code this directly in one line:

```
sqrt(sum((y - mean(y))^2 / (length(y) - 1)))

[1] 3.585686
```

Which we can confirm gives the same result as the **sd()** function:

```
sd(y)

[1] 3.585686
```

A.4   Model coefficients

Let's begin by creating making a linear model object in R called **check**. We will compare our manual results against those contained within this object later:

```
check <- lm(y ~ x, data = df)
```

A.4.1   Slope

The slope is of a linear model is calculated by dividing the sum of cross-deviations of y (SS_{xy}) by the sum of squares of x (SS_{xx}):

$$slope(\beta_1) = \frac{SS_{xy}}{SS_{xx}}$$

The equations for SS_{xy} and SS_{xx} are given below:

$$SS_{xy} = \sum xy - \frac{\sum x \sum y}{n}$$

and

$$SS_{xx} = \sum x^2 - \frac{(\sum x)^2}{n}$$

Step 1: Identify the variables/objects required: x, y, x^2, xy, $\sum x$, $\sum y$, $\sum x^2$, $\sum xy$, and n.

Step 2: Create any new variables/objects required: x^2, xy, $\sum x$, $\sum y$, $\sum x^2$, $\sum xy$, and n:

```
x2 <- df$x^2
xy <- df$x * df$y
sum.x <- sum(df$x)
sum.y <- sum(df$y)
sum.x2 <- sum(x2)
sum.xy <- sum(xy)
n <- nrow(df)
```

Step 3: Code the equation into manageable steps.

Calculate SS_{xy} then SS_{xy}:

```
SSxy <- sum.xy - sum.x * sum.y / n
SSxx <- sum.x2 - sum.x^2 / n        .
```

Calculate slope:

```
my.slope <- SSxy / SSxx
my.slope

[1] 0.9490539
```

Compare our value to that calculated by R by examining the second element of the coefficient of the model object using **coef()**:

```
coef(check)[2]

      x
0.9490539
```

A.4.2   Intercept

Once we know the slope we can use the following equation to calculate the intercept:

$$intercept(\beta_0) = \frac{\sum y - slope \sum x}{n}$$

Step 1: Identify the variables/objects required: $\sum y$, *slope*, and $\sum x$, *xy*, *n*.

Step 2: Create any new variables/objects required: none required.

Step 3: Code the equation into manageable steps:

Calculate the intercept:

```
my.intercept <- (sum.y - my.slope * sum.x) / n
my.intercept

[1] 0.2634643
```

Compare our value to that calculated by R object by examining the first element of the coefficient of the model object using **coef()**:

```
coef(check)[1]

(Intercept)
  0.2634643
```

A.4.3 r^2 (Pearson's correlation coefficient)

One method to calculate Pearson's correlation coefficient, r is:

$$r = \frac{SS_{xy}}{\sqrt{SS_{xx}SS_{yy}}}$$

As a result r^2 is simply:

$$r^2 = r^2$$

And SS_{yy} is:

$$SS_{yy} = \sum y^2 - \frac{\left(\sum y\right)^2}{n} \tag{A.1}$$

Step 1: Identify the variables/objects required: SS_{xx}, SS_{yy}, y^2, $\sum y$, $\sum y^2$, and SS_{yy}.

Step 2: Create any new variables/objects required: y^2, $\sum y^2$, and SS_{yy}:

```
y2 <- df$y^2
sum.y2 <- sum(y2)
SSyy <- sum.y2 - sum.y^2 / n
```

Step 3: Code the instructions into manageable steps:

```
r <- SSxy / sqrt(SSxx * SSyy)
r2 <- r^2
r2

[1] 0.859421
```

Compare our calculated value to that calculated by R:

```
summary(check)$r.squared

[1] 0.859421
```

A.4.4   AIC and AICc

AIC

The equation for AIC is:

$$AIC = 2K - 2L$$

where:

$$K = \text{number of calculated coefficients} + 1^*$$

and

$$L = \frac{-n}{2}\left(\log(2\pi) + 1 + \log\frac{RSS}{n}\right)$$

where

$$RSS = SS_{yy}(1 - r^2)$$

Step 1: Identify the variables/objects required: K, L, n, RSS, SS_{yy}, and r^2.
Step 2: Create any new variables/objects required: L, RSS, and K.

```
RSS <- SSyy * (1 - r2)
log.likelihood <- -n / 2 * (log(2 * pi) + 1 + log(RSS / n))
K <- length(check$coefficients) + 1
```

Step 3: Code the instructions into manageable steps:

```
my.AIC <- 2 * K - 2 * log.likelihood
my.AIC

[1] 32.37008
```

---

*representing a parameter for the calculation of the error term

Compare our calculated value to that modelled by the **AIC()** function:

```
AIC(check)

[1] 32.37008
```

AICc

The equation for the small sample size correction for AIC is:

$$AICc = AIC + \frac{2K(K+1)}{(n-K-1)}$$

Step 1: Identify the variables/objects required: AIC, n, and K.

Step 2: Create any new variables/objects required: none required.

Step 3: Code the instructions into manageable steps:

```
my.AICc <- my.AIC + 2 * K * (K + 1)/(n - K - 1)
my.AICc

[1] 38.37008
```

Compare our calculated value to that modelled by the **AICc()** function from the AICcmodavg package:

```
library(AICcmodavg)
AICc(check)

[1] 38.37008
```

Index

Printed in the United States
by Baker & Taylor Publisher Services